Concepts of Modern Biology Series

William D. McElroy and Carl P. Swanson, Editors

Animal Parasitism, *Clark P. Read*
Behavioral Aspects of Ecology, *Peter H. Klopfer*
Concepts in Bioenergetics, *Leonardo Peusner*
Concepts of Ecology, *Edward J. Kormondy*
Critical Variables in Differentiation, *Barbara E. Wright*
Euglenoid Flagellates, *Gordon F. Leedale*
Genetic Load: Its Biological and Conceptual Aspects, *Bruce Wallace*
An Introduction to the Structure of Biological Molecules,
   J. M. Barry and E. M. Barry
Molecular Biology: An Introduction to Chemical Genetics,
   J. M. Barry and E. M. Barry
The Organism as an Adaptive Control System, *John M. Reiner*
The Origins of Life on the Earth, *Stanley L. Miller and Leslie E. Orgel*
Processes of Organic Evolution, *G. Ledyard Stebbins*

Concepts of Modern Biology Series
William D. McElroy and Carl P. Swanson, Editors

CONCEPTS IN BIOENERGETICS

LEONARDO PEUSNER

Arthur D. Little, Inc.
Cambridge, Massachusetts

Prentice-Hall, Inc.
Englewood Cliffs, New Jersey

Library of Congress Cataloging in Publication Data

Peusner, Leonardo
   Concepts in bioenergetics.

   (Concepts of modern biology series)
    1. Bioenergetics. I. Title. [DNLM: 1. Bio-
physics. 2. Thermodynamics. QT34 P514c 1974]
QH510.P48   574.1'9121   73-15724
ISBN 0-13-166272-4
ISBN 0-13-166264-3 (pbk.)

© 1974 by Prentice-Hall, Inc., Englewood Cliffs, New Jersey

10 9 8 7 6 5 4 3 2 1

Printed in the United States of America

Prentice-Hall International, Inc., *London*
Prentice-Hall of Australia, Pty. Ltd., *Sydney*
Prentice-Hall of Canada, Ltd., *Toronto*
Prentice-Hall of India Private Limited, *New Delhi*
Prentice-Hall of Japan, Inc., *Tokyo*

CONTENTS

Preface   xiii

1   Energy, Thermodynamics, and Living Processes   **1**

1-1. Energetic nature of biological processes   *1*
1-2. What thermodynamics is about   *1*
1-3. The practical uses of thermodynamics in biology   *2*
1-4. Bird's eye view of the basic energetic processes of the biosphere   *3*
1-5. The dynamic design of plants and animals   *7*
1-6. How biological systems utilize their energy   *10*
1-7. Levels of biological organization   *10*
1-8. The large levels of energy utilization and organization: Ecosystems   *15*

2   The First Law of Thermodynamics   **19**

2-1. Thermodynamic systems   *19*
2-2. Examples   *21*
2-3. Equilibrium   *22*
2-4. Intensive vs. extensive variables   *23*
2-5. Not all variables are needed to specify the equilibrium states uniquely   *24*
2-6. Functions of state   *25*
2-7. Is there anything special about functions of state?   *25*
2-8. Cyclic processes   *26*

2-9. Some functions of state are defined in terms of
combinations of other quantities which are not
themselves functions of state   27

2-10. Energy is a function of state   27

2-11. Work   28

2-12. How work is calculated in specific cases   29

2-13. Work can be converted into kinetic energy   33

2-14. Is heat a fluid?   35

2-15. Metabolic heat and energy conservation   35

2-16. The mechanical equivalent of heat   36

2-17. Energy is a function of state   37

3   The Direction of Natural Processes: Entropy and Free Energy   41

3-1. The Second Law of Thermodynamics   42

3-2. Reversible vs. irreversible processes   42

3-3. Entropy   43

3-4. Entropy changes during irreversible processes   46

3-5. Entropy is a function of state!   47

3-6. Entropy changes and work   48

3-7. The First and Second Laws combined   49

3-8. Free energy: The biological function of state   49

3-9. The useful work and ΔG 51

3-10. The explicit expression for the free-energy change   52

3-11. Example   53

3-12. Can mass ever move from a low potential to a high
potential?   56

3-13. Explicit form of the chemical potential   56

3-14. Complete form of the chemical potential   59

3-15. Ions and the electrochemical potential: 60

3-16. The electrochemical potential   62

4   Microscopic Interpretation of Thermodynamic Quantities   65

4-1. But what is entropy?   65

4-2. Ideality and molecular cohesion   66

4-3. Ideal gas kinetics   67

4-4. The probabilistic nature of entropy   70

4-5. Entropy, probability, order, and disorder   72

4-6. Relationship between microscopic energy and
microscopic entropy   *73*

4-7. Energy, temperature, and pressure of the ideal gas   *76*

4-8. The Second Law and the quality of energy   *79*

5 Free Energy in Molecular Biology and Bioenergetics   81

5-1. Relationship between standard free-energy changes
and equilibrium constants   *81*

5-2. Difference between direction and rate   *84*

5-3. Covalent vs. weak bonding: Molecular biology as the
chemistry of molecular recognition   *86*

5-4. What is the physical basis of weak interactions?   *89*

5-5. Is the hydrogen atom charged?   *91*

5-6. Activation energies of weak bonds   *91*

5-7. If weak bonds form and break so fast, are there a
substantial number present at equilibrium?   *92*

5-8. The importance of three-dimensional fitness   *92*

5-9. Three-dimensional shapes of macromolecules are
configurations of minimum free energy   *94*

5-10. How biological free energy is obtained and put to
work: Bioenergetics   *94*

5-11. The recovery of chemical energy   *97*

5-12. Coupling chemical reactions   *98*

5-13. Other forms of biological work also require ATP
hydrolysis   *102*

5-14. Conservation of biological energy: The generation of
ATP   *102*

**6 The Electrochemical Potential as a Measure of Biological Equilibrium:
Osmotic Pressure, Nernst and Donnan Potentials   107**

6-1. When is the electrochemical potential a measure of
equilibrium?   *107*

6-2. Biological resting potentials   *108*

6-3. Quantitative derivation of the electrical potential:
Nernst Equation   *110*

6-4. Microscopic separation of charge   *112*

6-5. Action potentials   *114*

6-6. What came first: Potentials or concentration
differences?   *116*

6-7. The importance of equilibrium calculations; activity coefficients   *117*

6-8. Are activities arbitrary quantities? The glass electrode   *118*

6-9. Gibbs-Donnan equilibrium   *119*

6-10. Gibbs-Donnan equilibrium in the erythrocyte   *121*

6-11. Chemical potential and osmotic equilibrium   *122*

6-12. Physical meaning and biological implications of osmotic pressure   *123*

6-13. Turgor pressure and stomatal opening   *125*

6-14. Osmosis is a very strong force   *129*

6-15. Colloid osmotic pressure and the motion of water in capillaries   *130*

6-16. Reflection coefficients   *131*

6-17. A possible way to measure σ   *133*

6-18. Osmotic pressure in the presence of several solutes   *134*

6-19. Osmotic regulation and habitat   *134*

7 Chemical Potential in Action: Diffusion   139

7-1. The meaning of the chemical potential away from equilibrium   *139*

7-2. Visualization of the potential   *141*

7-3. Steady velocity   *141*

7-4. Explicit expression for steady flow   *143*

7-5. Fick's Law   *144*

7-6. Physical meaning of Fick's Law   *145*

7-7. Intuitive meaning of diffusion   *146*

7-8. Einstein's molecular view of diffusion   *147*

7-9. Evolutionary pressures imposed by diffusional processes: Diffusion and the size of organisms   *150*

7-10. Why are cells so large?   *153*

7-11. The problem of fluctuations   *153*

7-12. Gas diffusion   *156*

7-13. Respiration in vertebrates as an example of gas diffusion   *157*

7-14. Some unrelated transport processes   *159*

7-15. The nephron   *162*

7-16. Poiseulle, or bulk, flow: Pores in membranes   *164*

7-17. Applications to biological membranes   *165*

7-18. Interaction between osmotic and hydrostatic gradients   *165*

7-19. In dilute solutions J_v is the velocity of the solvent   *167*
7-20. Hydrostatic pressures "push" solute molecules   *168*
7-21. Diffusional flow in the presence of hydrostatic and
concentration gradients   *169*

8 Information Theory, Codes, and Messages   177

8-1. Information theory   *177*
8-2. The primitive telegraph   *178*
8-3. The genetic message, a biological example   *178*
8-4. Efficient coding   *182*
8-5. Optimum man-made codes; binary system   *182*
8-6. Probability and information content   *186*
8-7. Encoding and decoding redundant messages   *188*
8-8. Biological mistakes   *189*
8-9. Error-correcting codes   *190*
8-10. Information theory and thermodynamics; Maxwell's
demon   *192*
8-11. Information content and entropy   *193*
8-12. Reversible vs. irreversible transmitters   *193*
8-13. Information and knowledge   *195*

9 Information Theory and Biology   197

9-1. How many bits can the human brain store?   *197*
9-2. Is biological memory stored in informational
macromolecules?   *198*
9-3. Dancing bees and chemical ants   *200*
9-4. The information content of a bacterial cell   *204*
9-5. Functional number of genetic messages   *205*
9-6. The emergence of life as an accident   *207*
9-7. How did life begin?   *208*
9-8. Life in other worlds?   *210*
9-9. Evolution of proteins   *212*
9-10. Capacity of a channel: Crayfish photoreceptor   *214*
9-11. Is this coding efficient?   *217*

10 Thermodynamic Efficiency, Biological and Mechanical Machines   219

10-1. Efficiency of heat engines   *219*
10-2. Two obvious problems   *221*

10-3. The way around the equilibrium problem: Local
      equilibrium and steady state   *222*

10-4. Energy vs. power   *224*

10-5. Transducers   *225*

10-6. Biological processes consist of many transducers
      attached to one another   *227*

10-7. How can transducers be described in quantitative
      terms?   *228*

10-8. Dynamic efficiency and Onsager (nonequilibrium)
      thermodynamics   *228*

10-9. Spontaneous coupling of two processes requires
      positive entropy production   *230*

10-10. How nonequilibrium thermodynamics changes our
       perception of some biological problems   *230*

10-11. Coupling and asymmetry   *231*

11 Thermodynamic Efficiency and Ecology   235

11-1. Efficiency of energy flow in the biosphere: Implications
      from the Second Law of Thermodynamics   *235*

11-2. What are the relative efficiencies of producers and
      consumers?   *239*

11-3. Pollution and the smogmobile   *240*

11-4. Pollution and thermodynamics   *241*

11-5. Second Law and pollution   *241*

11-6. On technical, conversion, and application
      efficiencies   *242*

11-7. Gas vs. electric heat   *244*

11-8. Power vs. efficiency   *244*

11-9. Can pollution be completely avoided?   *247*

11-10. How low can you get?   *250*

11-11. The "principle" of minimum entropy production   *252*

**APPENDIX Mathematical Concepts in Thermostatics and Nonequilibrium
           Thermodynamics   255**

A-1. Reversible work   *255*

A-2. Exact differentials and functions of state   *256*

A-3. First and Second Laws of Thermodynamics   *258*

A-4. Other energylike state functions   *259*

A-5. Explicit form of the chemical potential   *260*

A-6. Differential definition of the chemical
potential   *261*
A-7. The electrochemical potential   *261*
A-8. Chemical potentials in equilibrium and away from
equilibrium   *262*
A-9. External forces and steady state   *263*
A-10. Chemical potential outside equilibrium: Fick's
Law   *264*
A-11. Chemical reactions in the steady state   *265*
A-12. The internal production of entropy   *266*
A-13. Setting up the proper equations   *268*
A-14. Diffusion of a solute in water   *268*
A-15. Passage of solute and water across a
membrane   *270*

Glossary   273

Solutions to Problems   277

Author Index   299

Subject Index   301

PREFACE

This book originated in the form of a series of notes in advanced biological thermodynamics I provided to Harvard undergraduates as a complement to my lectures in a bioenergetics course.

In the process of reworking this manuscript, however, I attempted to broaden the treatment of biological concepts and at the same time make them useful also to students with less advanced mathematical background. To this end, calculus has been eliminated from the body of the text and relegated to the appendix, which provides detailed mathematical treatment. I believe this has made the material more generally applicable to such fields as physiology, molecular biology, ecology, and biophysical physiology, and it may be of interest to workers in these fields as well as to the original audience of premedical and biological sciences students. Wherever possible, I have included problems and examples to show the application of the concepts introduced in the text; solutions to these problems are provided at the end of the book.

I would like to acknowledge the valuable secretarial help given, at various times, by Linda Dolmatch, Dorothy Dickenson, Betsey Cobb, Virginia Doty, and particularly Barbara Arimento. My friend Thomas Strunk gave me early encouragement when the project looked rather improbable. My thanks to Peter Curran, S. Roy Caplan, Donald Mickulecky, and Stanley Schultz, who first introduced me to biological thermodynamics, and to Arthur K. Solomon for giving me the opportunity to teach bioenergetics to Harvard undergraduates for several years. Last but not least I wish to thank my wife Kenna for carefully reading the manuscript and final proof, and the Prentice-Hall production editor, Zita de Schauensee, who efficiently directed the assembly of the book in its various stages of production.

Cambridge, Massachusetts   *Leonardo Peusner*

1

ENERGY, THERMODYNAMICS, AND LIVING PROCESSES

1-1. Energetic Nature of Biological Processes

The essential feature of living organisms is their ability to capture, transform, and store various forms of energy according to the specific instructions carried by their individual genetic materials. On a larger scale, life on earth—its origin, evolution, and adaptation to the physical conditions of our planet—depends on the efficient trapping of the energy of our star, the sun, and its allocation into various biosynthetic processes.

The objective physicochemical explanation of how life works, started, and evolved must therefore center around the laws that dictate the way in which living systems acquire, degrade, and organize energy. This is the goal of a relatively young discipline, biological thermodynamics, which brings together the results of classical thermodynamics and information theory and applies them to the problems presented by biological systems.

1-2. What Thermodynamics Is About

Thermodynamics deals with the general laws that regulate energy transactions. Usually, thermodynamics considers "engines" that transform one kind of energy into another kind of energy; the engines of thermodynamics are not, however, restricted to man-made machines or to machines with a special structure or design: The laws dealt with in thermodynamics have universal validity, and they hold no matter where the given process takes place.

Originally, thermodynamics dealt with man-made machines: The main preoccupation of early thermodynamicists was to improve the efficiency of the steam engine that had brought about the industrial revolution. Thermodynamics was then directly concerned with an effort to replace man's work with inanimate objects that could work faster and perform more work than he. Although the precise applications to biology are fairly recent, thermodynamics has helped to eliminate the artificial distinction between living and inanimate things.

Is thermodynamics hard? Yes and no. Thermodynamics rests on two simple laws. One makes a statement about the conservation of energy and the other specifies the direction of natural processes. The mathematical foundations of thermodynamics, however, can be involved, as any person who has taken an introductory course in physical chemistry can testify. This monograph takes the unorthodox approach of eliminating all differential calculus from the text, which, in my opinion, makes life easier. Mathematical principles and the more advanced irreversible thermodynamics are given in the Appendix.

1-3. The Practical Uses of Thermodynamics in Biology

If the only purpose of biological thermodynamics were to demonstrate that living things follow physical laws, there would be little reason to study the subject. The main importance of thermodynamics is that it can provide information as to how a given biological system may be working and suggest logical experiments to test a given theory.

Unfortunately, many biologists view with suspicion the use of physics in their field. The most important reason for the complex interaction between experts in physics and in biology is a deep-rooted prejudice on both sides: while the physicist is convinced that a biological system can always be described in mathematical terms, many biologists still believe that life is driven by a nonmaterial principle, the "vital force," which is different from the energy found in the inanimate world.

Both extreme positions have had their champions. Louis Pasteur's experiments on spontaneous generation convinced him that life had an eternal character and that as such it had preceded matter in the universe. On the other hand, Erwin Schrödinger, one of the founders of quantum theory, remarked that living systems studied by physical means must always obey the laws of physics. However true that may be, this last statement contains the implicit assumption that given *any* biological phenomenon, it is always possible to find a formula or sets of formulas representing the event in physical terms.

As usual, practical truth is somewhere between the two extremes. While recent progress in biology has brought experimental observation closer to physical science than ever before, biophysics cannot predict and describe completely the behavior of the biological world.

1-4. Bird's Eye View of the Basic Energetic Processes of the Biosphere

Although it is not the purpose of this book to teach general biology, we review the general scheme of energy acquisition and utilization by the biosphere to place our physical examples in the biological framework.

All the energy that drives life on earth comes from one of the most destructive forces known to man: a hydrogen bomb. In the sun, hydrogen is continuously being transmuted into helium, and as 4 atoms of hydrogen fuse to form a single atom of helium, some mass "vanishes" (0.029 gram every time 4.003 grams of He are formed, to be precise) and transforms into energy according to the overpopularized formula of Einstein, $E = Mc^2$, in which M is the mass that disappears and c is the velocity of light, 300,000 meters per second. Since over a hundred million tons of mass disappear from the sun every minute and since the velocity of light is so great, large amounts of energy are generated. This excess energy does not remain in the sun but is radiated into space in the form of electromagnetic waves—light, radio waves, X rays, heat, etc. (Fig. 1-1). This process of conversion of mass into energy can take place in the sun because of the extreme temperature conditions but it does not occur under normal circumstances on the surface of our planet.

How does the biosphere detect the energy of the sun? Although electromagnetic waves are continuously around us, we can see only a small portion of the solar spectrum, the part we call visible light and the part we detect as "heat." The rest of the spectrum is invisible as far as we are concerned. Visibility, however, is not a property of the electromagnetic wave itself but of the organism on which it impinges. The energy carried by T.V. signals, for example, is invisible to humans; but when an appropriate receptor is placed in the pathway of these signals, they can be converted into images we perceive. Furthermore, different stations transmit signals of different frequencies; it is therefore necessary to tune in to a given station in

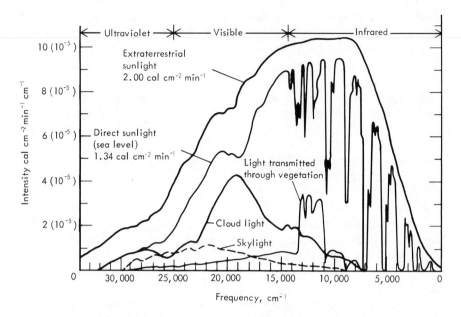

Fig. 1-1. The intensity of the sunlight reaching the earth is not uniform but depends on the frequency, or color, of the light. In addition, clouds and vegetation act as filters and transmit certain wavelengths preferentially. (Reproduced from D. M. Gates, "The Energy Environment in Which We Live," *American Scientist*, vol. 51, p. 327.)

order to "capture" the signals it sends. Similarly, the human eye is tuned to the "visible" spectrum—the part of the spectrum that includes the colors of the rainbow. The radiation of the sun covers the whole spectrum, but only a small portion of it actually reaches the earth. Large amounts of the energy arriving on the earth are directly reflected by the atmosphere; this is the ultraviolet part of the spectrum, and 30% of the arriving energy is lost this way (a fortunate circumstance since U.V. radiation is extremely dangerous). About 50% of the coming energy is converted into heat and reradiated into outer space as infrared radiation, and 20% is utilized to evaporate water and form clouds. Finally, only about 0.02% of the incoming energy is detected by the ecosphere (Fig. 1-2)!

The actual detection of solar energy is achieved by the green plants, which are the "tuners" of the biological world. While animals cannot utilize the energy detected from the sun, green plants trap the energy of the sun's light and manufacture food substances that serve as the "energy pills" of the biological world. The process by which green plants capture this energy and store it into sugars, the most

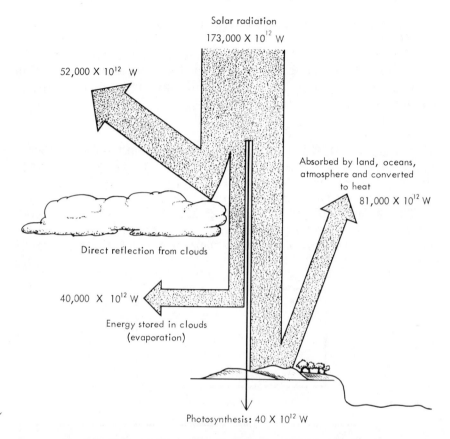

Fig. 1-2. Only a small amount of the total solar energy reaching the earth is fixed by photosynthesis. The thickness of the arrows represents the amount of energy absorbed, reflected, or stored per unit time in units of watts (W). (The U.S. energy consumption per unit time is approximately 3×10^{12} W.)

basic chemical energy source, is called *photosynthesis.* Photosynthesis is represented by the equation

$$H_2O + CO_2 + h\nu \longrightarrow C_6H_{12}O_6 + O_2$$

(water + carbon dioxide + energy → glucose + oxygen).

The actual detector is a pigment or colored molecule, chlorophyll, which is also responsible for the green color of plants. The whole functioning of the biosphere is dependent upon the efficient detection and trapping of solar energy by chlorophyll: since animals cannot perform photosynthesis, they must acquire their fuel either by eating plants (herbivores) or other animals that feed on plants (carnivores).

The basic process by which both higher animals and plants use the energy stored in sugars and other food molecules consists in burning these molecules in the presence of oxygen during respiration, which is represented by the equation

$$O_2 + food \longrightarrow H_2O + CO_2 + energy.$$

When sugar is burned in the air, energy is evolved in the form of heat. Similarly, when an organism burns foodstuffs, energy is evolved; part of it is stored in a chemical form and finally released to perform work. The basic scheme is shown in Fig. 1-3.

The design of plants and higher animals can be easily understood in terms of these two processes. There are, however, some organisms that do not need oxygen to obtain their energy and also some plants

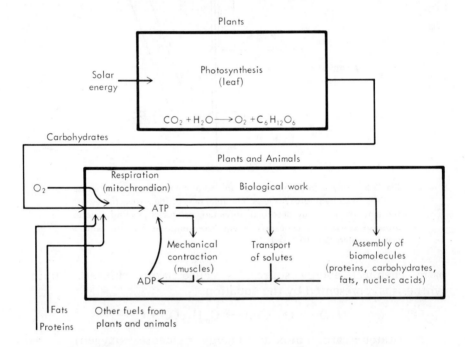

Fig. 1-3. Generation and utilization of chemical energy in common biological systems. The chemical energy contained in foods—carbohydrates, fats, and proteins—is released and re-stored in the form of an "energy rich" substance, adenosine triphosphate (ATP), which can release its energy for use in mechanical, transport, or chemical work. Whereas carbohydrates, proteins, and fats serve for long-term energy storage, ATP is a short-term energy-storing molecule.

that utilize pigments different from chlorophyll to detect the energy of the sun. Moreover, some microorganisms can obtain the required energy by utilizing inorganic molecules.

1-5. The Dynamic Design of Plants and Animals

Since photosynthesis is the main activity of green plants, it is not surprising to find that the main structure of the green plant, the leaf, is the seat of photosynthesis. Leaves provide support for chlorophyll molecules, large surfaces for capturing the light arriving from the sun, and openings called *stomata* for the exchange of oxygen and carbon dioxide. Water required for photosynthesis arrives through a "plumbing" system called *xylem*, which carries salts, while the sugar produced is exported to every part of the plant through another plumbing system called *phloem*. Xylem and phloem constitute the circulatory system of the typical plant (Fig. 1-4).

The other basic functional structures are the *root*, which absorbs minerals and water from the soil and also provides support for the whole plant, and the *stem*, which has the dual role of supporting branches, leaves, and flowers and serving as a continuous passage for phloem and xylem from the root to the leaves. *Flowers*, incidentally, are transformed leaves which constitute the sexual organs of flowering plants.

Just as the organization of a typical plant is geared toward the efficient conversion of electromagnetic energy into chemical energy (food), so the typical higher animal is designed for the efficient burning of foodstuffs in the presence of oxygen. Since the actual "burning" of fuel takes place in every part of the animal or plant, it is important to have, in large animals, a system that distributes oxygen and fuels throughout the organism. In very small, primitive animals such a system is not required because each part of the animal is in close contact with an oxygen source (either in the air or dissolved in water); in higher vertebrates, including man, there is a circulatory system that carries gases and food to and from every part of the body. The blood acquires oxygen and liberates carbon dioxide at the lungs, and it incorporates foodstuffs as it passes by the intestine (Fig. 1-5). The processes of digestion and mastication are directed at reducing the complex constituents of food into simpler substances which can easily be assimilated at the intestinal level and incorporated into the blood. Food utilization and energy recovery by the animal or plant is called *metabolism*. As a result of the metabolic activities, noxious products appear in every part of the organism;

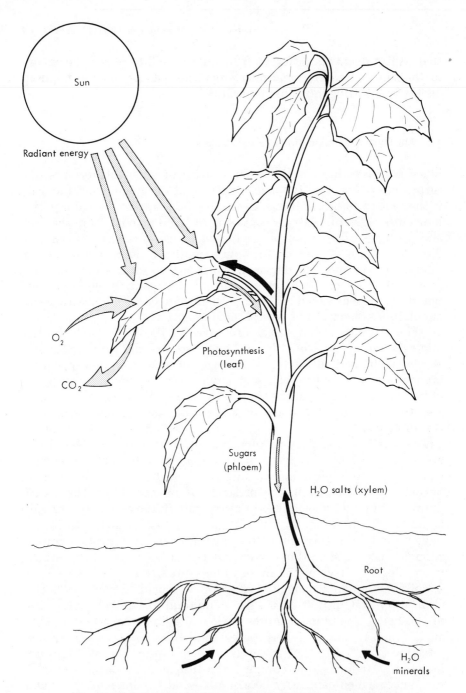

Fig. 1-4. Organization of a flowering plant. The basic organ is the leaf, which traps radiant energy and eventually transforms it into chemical form through the process of photosynthesis. The sugars synthesized during photosynthesis are long-term energy-storing molecules.

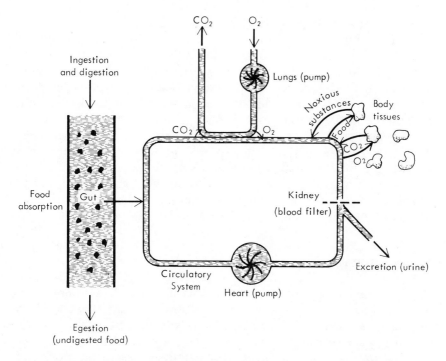

Fig. 1-5. A simplified view of the basic transport activities that provide energy to the tissues of a vertebrate. Foodstuffs are absorbed in the small intestine and carried to every cell of the body, where the actual energy recovery takes place. The oxygen required for respiration, as well as the resultant CO_2 and metabolic by-products, is also carried by the blood. The kidney acts as a continuous filter, removing noxious substances from the circulatory system. During these various activities gases and solids must cross cellular and other boundaries. In some cases the passage is spontaneous; in others metabolic chemical energy is required. (Adapted from C. L. Prosser et al. *Comparative Animal Physiology*. Philadelphia: W. B. Saunders and Co., 1950, p. 761.

these products are also sent to the blood, so it becomes necessary to filter them out. This job is done by the kidney, which sends noxious materials to the urinary bladder for future excretion from the body. Other organs (liver, pancreas, gallbladder) also take part in the process of energy recovery and storage, but we limit our overall view to the basic, general points.

Most animals cannot acquire energy simply by waiting for the sun to fall on their heads: Since they lack the photosynthetic equipment, they must look for their food and get it either from plants or from other animals that ingest plants. Some animals, of course, are parasites and get their food from a host, but these are the exception rather than the rule. Animals, then, must possess a special system to move their wings, legs, or fins: the muscular system. Animals also need detectors, collectively known as the sensory system, to find their food. Sensory systems range from eyes, nose, and ears to the bizarre design of the bat's sonar—a system similar to that used by British ships in the Second World War to alert them to the presence of German submarines. The bat, however, uses its sensory system to catch moths. Sensory detectors are the peripheral expression of the central nervous system, which has the role of providing a rapidly adapting system (it can respond fast to external stimuli). Animals as well as plants also possess long-term regulatory systems: These are the chemical regulators, or hormones, that are secreted by internal glands (Fig. 1-6).

1-6. How Biological Systems Utilize Their Energy

Organisms acquire energy to do biological work; large amounts of energy, however, are wasted in the form of heat. What kinds of biological work must be done? Some of the obvious places where an organism is seen to do work can be detected from the system drawing of Fig. 1-5. The heart, for example, has to do work as it pumps the blood. Less obvious, perhaps, but equally important is the work done by the intestine to absorb foodstuffs. Work must also be done to "raise" the energy-storing molecules from their low-energy to their high-energy form, as shown in Fig. 1-3.

Work is also done at the high level of genetic and nervous integration. In order to function effectively, biological systems must contain programs on how their internal machinery should function, and they must acquire information about the internal and external environment. This type of work involves organizing symbols, which in turn control energetic processes, organize biostructure, or control the energy needed for fast response or to generate associations of other symbols.

1-7. Levels of Biological Organization

All tubes, pumps, filtering systems, etc., that constitute the complex machinery of Fig. 1-5 are formed by specialized aggregations of the same basic unit, the cell. The cell is the minimal structure that can be considered to be alive in the same way that the atom is the minimal structure that has the physicochemical properties of an element. All

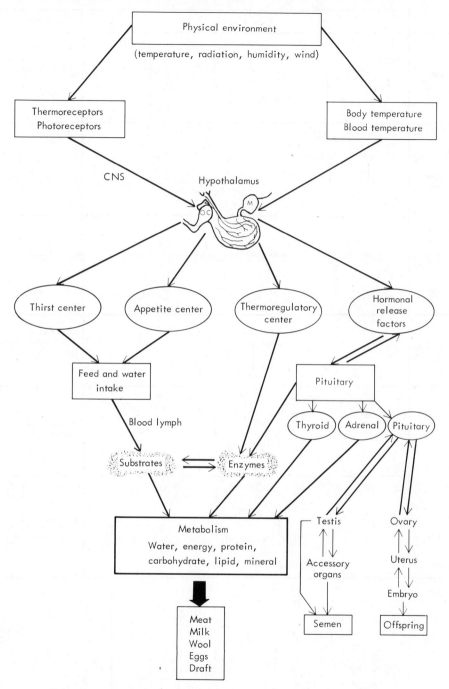

Fig. 1-6. Animal productivity is controlled by the central nervous system and hormonal mechanisms. External energy and metabolic energy are required to effect these higher control activities. (Reproduced from E. S. E. Hafez, *Adaptations of Domestic Animals*. Philadelphia: Lea and Febiger, 1968, p. 91.)

	Structural formula	Comments
Sugars (monosaccharides)	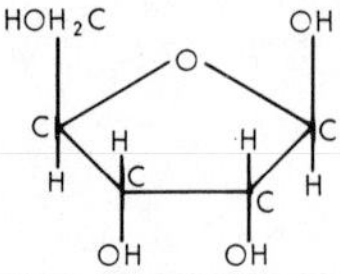 **Deoxyribose** 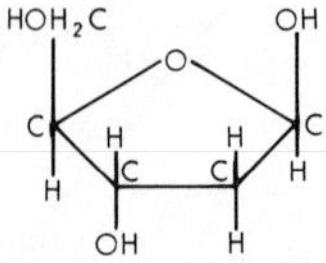 HOH_2C OH C C H H OH H — **Ribose** HOH_2C OH C C H H OH OH — **Glucose** HOH_2C H C H H OH H C OH HO OH 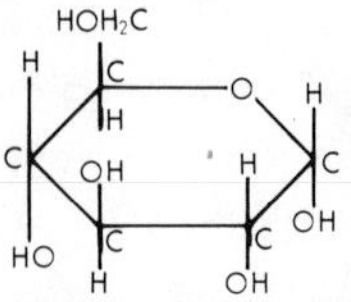	Monosaccharides (also called carbohydrates) usually obey the general formula $C_n(H_2O)n$; e.g., ribose has $n=5$, glucose $n=6$.
Purines	**Adenine** 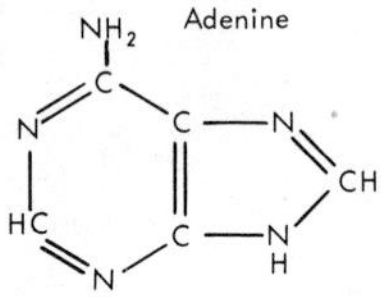 NH_2 C N CH HC C N N H — **Guanine** O C HN C N CH C C N H_2N N H 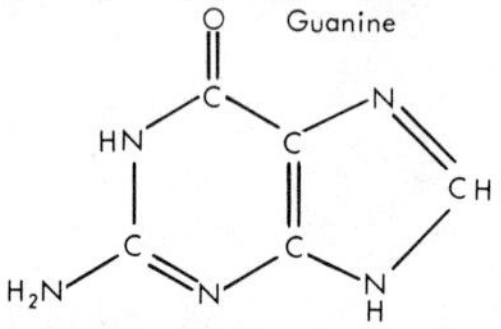	
Pyrimidines	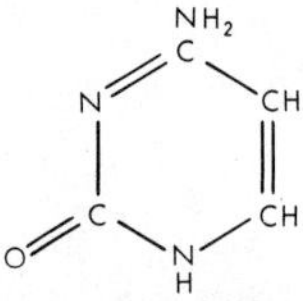 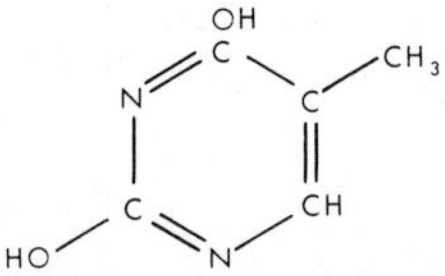 **Cytosine** NH_2 C N CH C O C N CH H — **Thymine** OH C N CH_3 C C CH HO N — **Uracil** O C HN C H C O C N H 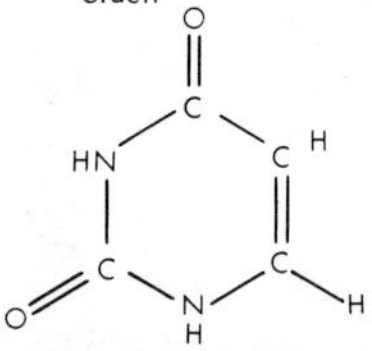	
Amino acids	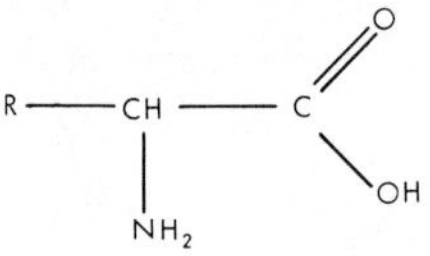 R—CH—C=O, OH, NH_2	R is any one of 20 possible organic groups; e.g., CH_3, H, C_3H_7, CH_2OH, etc.
Nucleotides	**DNA nucleotides** Phosphate 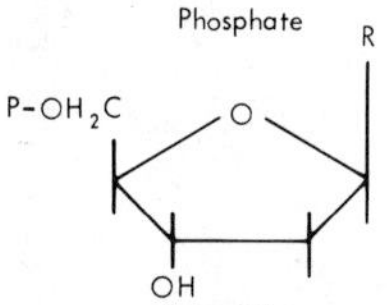 $P-OH_2C$ O R Purine or pyrimidine base OH Deoxyribose	R is adenine, guanine, cytosine, or thymine.
	RNA nucleotides Phosphate 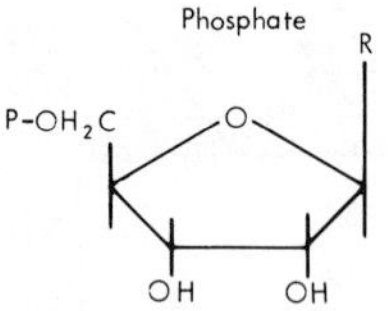 $P-OH_2C$ O R Purine or pyrimidine base OH OH Ribose	R is adenine, uracil, guanine, or cytosine.
Lipids	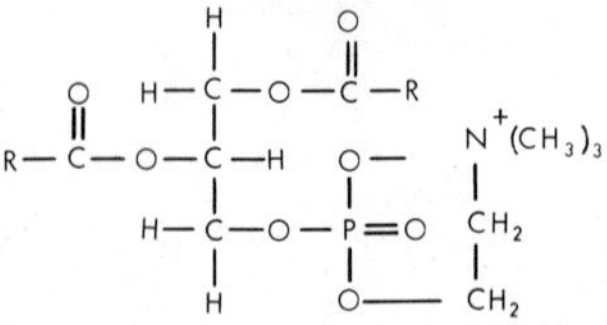 **Steroids** 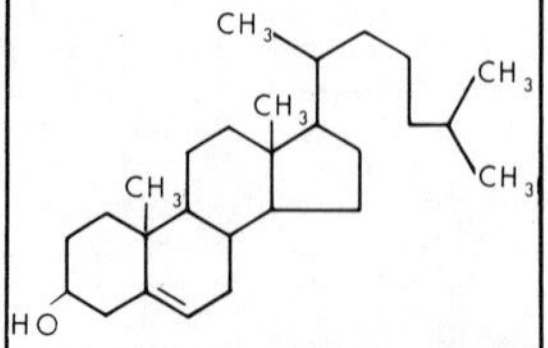 CH_3 CH_3 CH_3 CH_3 CH_3 HO Cholesterol — **Phospholipids** H, O, H—C—O—C—R, O, R—C—O—C—H, $N^+(CH_3)_3$, H—C—O—P=O, CH_2, H, O—, CH_2 Lecithin — **Fats** H, O, H—C—O—C—$(CH_2)_mCH_3$, O, H—C—O—C—$(CH_2)_nCH_3$, O, H—C—O—C—$(CH_2)_5CH_3$, H 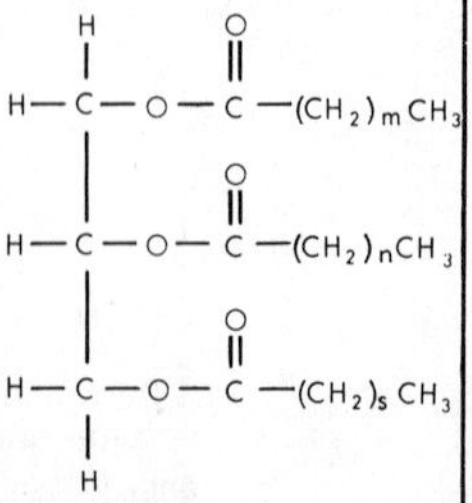	

Fig. 1-7. Basic small molecules that are characteristic of living systems.

12

cells have the same basic set of molecules, given in Fig. 1-7, but knowledge of these molecules alone does not define life; it is the *grouping* of these molecules that constitutes the living entity.

Cells are composed of a set of structures (each of which consists of many molecules) called *organelles*. Each organelle has a specific function; from the energetic point of view, the following are the most relevant (refer to Fig. 1-8):

- The *mitochondrion*, "power house" of the cell, transforms food energy so that it can be used by the cell.

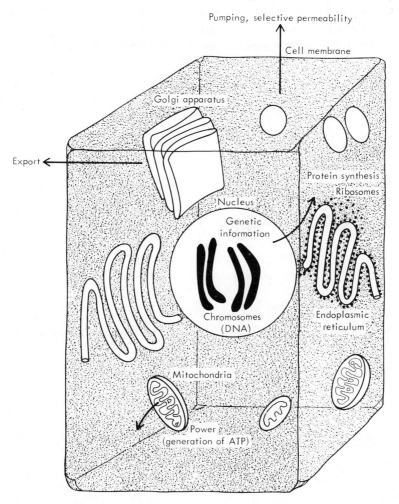

Fig. 1-8. A schematic diagram showing some of the important cellular components and their activities.

- In plant cells, *chloroplasts* contain chlorophyll, the pigment that detects energy coming from the sun.
- The *nucleus* contains chromosomes, which carry the genetic information necessary for specifying all the structure and function of the cell.
- *Ribosomes* have the primary role of reading the genetic message sent to the cytoplasm (any region in the cell that is outside the nucleus) and translating it into structure or function.
- The *cell membrane* serves to separate the internal from the external, unfavorable environment and to maintain, through transport of ions and metabolites, an appropriate environment for the chemical reactions for life.

Depending on the specific job a given cell must perform, one or more of the organelles, as well as other specializations, may be

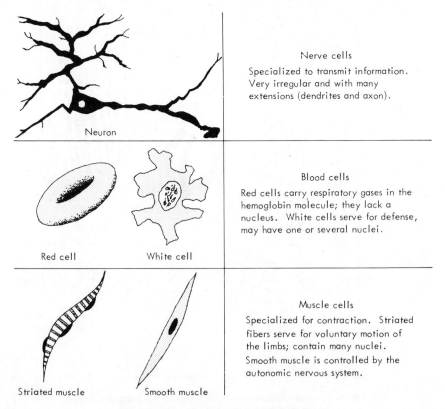

Fig. 1-9. Some typical cellular specializations.

predominant (Fig. 1-9). Sperm cells, for example, must transfer information through a relatively long distance; as a result, most of the cell is reduced to the chromosomes, in the head of the sperm, and a tail which serves for propulsion. Since large amounts of energy are needed, there are many mitochondria present in the sperm's tail. On another extreme we have red blood cells, whose job is to carry large amounts of oxygen and carbon dioxide. The mature red blood cell lacks most of the common organelles, including the nucleus in humans, and has many packed molecules of hemoglobin, which serve to carry respiratory gases. Cells in which large exchanges of matter take place increase their area by providing infoldings in the cellular membrane. Palisade cells in the leaf are packed with chloroplasts, as they have the burden of photosynthesis.

Just as small molecules associate to form larger molecules, organelles, and cells, so cells associate to form tissues (a group of similar cells which performs a general type of job, such as contraction, in muscle), tissues associate to form organs, which have a specific job such as pumping blood (heart), performing photosynthesis (leaf). Organs in turn associate to form systems, such as the digestive system and the respiratory system. Systems associate to form organisms, which associate to form communities and ecosystems.

1-8. The Large Levels of Energy Utilization and Organization: Ecosystems

The energetic requirements of each organism are not satisfied individually but through the establishment of a delicate equilibrium between the various organisms that live in a given region in a specific environment. There is a continuous flow of energy and matter between neighboring plants, animals, microorganisms, and the physicochemical environment. The sum total of all the organisms and the physical environment taking part in these energy transactions is called an *ecosystem*.

Superficially, ecosystems look very different one from another: The inhabitants of a lake, for example, are different from the inhabitants of a mountain. Clearly, the description of a given ecosystem becomes laborious if no unifying concepts are adopted; it is then necessary to catalog all the species of plants and animals present in a given ecosystem—such as a lake or a forest—together with all the exchanges of mass and energy that take place. The description of an ecosystem is enormously simplified when it is given in terms of the *types* of energy and mass transactions.

In any ecosystem there are two basic types of organisms: *producers* and *consumers*. The self-feeding organisms that can fix the energy from the sun and transform the materials from the immediate inanimate environment into food (fuel) are called *autotrophs*. Some autotrophs are microorganisms that can get both the energy and the matter needed from products found in the environment; others are microorganisms that can utilize products of decomposition of larger organisms to obtain their energy and mass. Organisms that cannot make their own food are called *heterotrophs*; these depend on the producers for their energy source.

The most important autotrophs are, as we have already pointed out, the green plants, which capture the energy of the sun through photosynthesis. There are, however, a few bacterial types that are able to derive energy not from the sun but from inorganic materials—from substances that would be present on earth even if life did not exist. These are *chemosynthetic* types. Although organisms that are self-feeding are called producers, we must keep in mind that they are not producing energy; they merely transform the energy coming from the sun (radiant energy) into chemical form.

All the elements that take part in building organisms are used over and over by the biosphere. The earth does not receive or lose significant quantities of mass from or to outer space, and there is no spontaneous interconversion of energy into mass on the earth. Mass that takes part in biochemical reactions, then, has come from, and is used by, other organisms. Claude Villee points out that the mass of all the life that ever existed on earth is probably several times the weight of the earth itself! Thus, recycling is imperative. Some of the atoms that form the reader's body may have been at one time or other in a plant, a dinosaur, a whale, or in human beings who lived in prehistoric times.

Which elements need recycling? The basic elements that form all living organisms—carbon, hydrogen, oxygen, nitrogen, phosphorous, and sulfur—are used in relatively large amounts, so it is mandatory that they be available at all times. If the supply of any one of these elements were suddenly stopped, life itself would stop. Generally, the situation is not so serious. The cycling process consists of the exchange of elements between the biological and the inanimate worlds; the exchanges are called *biogeochemical cycles*. They can be classified into more or less perfect cycles depending on whether the rate of transfer from the biological to the inanimate world is more or less equal to the transfer in the opposite direction.

Carbon has a perfect cycle: This means that it enters and leaves the biosphere (plants and animals) at the same rates. The total amount of carbon that is present in the biological world is relatively constant and, as a result, the total biomass does not change appreciably in time. Some large quantities of carbon are stored in the earth in the form of fossil fuels and are being released by men, with the possible result that once the amount of carbon in the form of CO_2 increases in the atmosphere it can increase the temperature of the earth, leading to a larger photosynthesis rate and larger plants. Elements that are needed in smaller quantities, such as sulfur and phosphorous, are actually more important in determining the stability of the biosphere, because small changes in their supply, especially deficiencies, can lead to large changes in the population of an ecosystem.

The flow of energy in ecosystems follows a different path from that of mass. While energy is transferred, like mass, from producers to herbivores, carnivores, and decomposers, it is not recycled but lost into outer space. Thus, the continuous input of energy from the sun is needed to maintain life on earth. The reason why energy is not reused has to do with a general tendency of energy degradation which exists throughout the universe and makes energy more and more unavailable as time progresses. This is one of the important thermodynamic concepts we shall study in detail. The general flows of mass and energy in an arbitrary ecosystem are shown in Fig. 1-10.

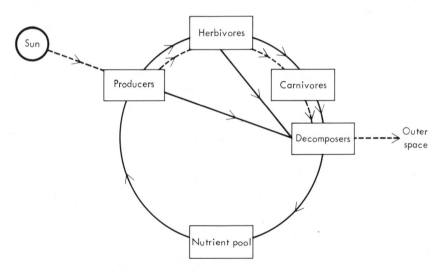

Fig. 1-10. Diagram showing energy and mass movement in the ecosphere. Dashed arrows indicate the flow of energy; black arrows, the flow of matter. It can be seen that matter is reused and follows a cyclic pattern, while the flow of energy is acyclic. (Adapted from E. J. Kormondy, *Concepts of Ecology*. Englewood Cliffs, N.J.: Prentice-Hall, Inc., 1969, p.4.)

2

THE FIRST LAW
OF THERMODYNAMICS

This chapter introduces the law of energy conservation, a basic rule of Nature which applies to biological as well as inanimate systems. The main step in the direction of stating this law consists in recognizing that heat and work are different expressions of the physical entity we call energy.

In order to state the principle of energy conservation, we must first introduce two simplifications. First, we subdivide the biological world into small regions that are simple to describe but encompass meaningful processes. Second, we assume that a static situation prevails. Both assumptions are weak from the biological point of view since equilibrium is attained only in the static case we call death and since the sum of parts rarely behaves as the complete biological entity. This initial approach is justified, however, because the dynamic nature of living things can be introduced with minimal formal changes later.

2-1. Thermodynamic Systems

A system is any part of the world we wish to observe and describe. Examples of relevant thermodynamic systems are: a cell, a mitochondrion, a heart, a rat, the earth. We usually draw an imaginary surface around the system and ask the following questions about the boundary obtained:

- Can matter pass through the boundary?
- Can heat pass through it?

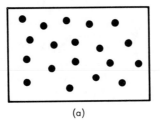

(a)

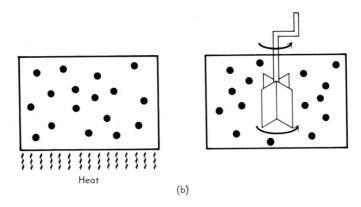

Heat

(b)

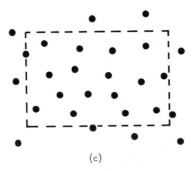

(c)

Fig. 2-1. (*a*) Isolated system: Neither matter nor energy can enter or leave the system. (*b*) Closed systems: Matter cannot enter or leave the system but energy can be introduced by doing work or adding heat. An *adiabatic* system is a special closed system that does not allow heat to enter or leave. A *diathermal* system is a closed system that allows heat to pass through the boundaries. (*c*) Open system: Both matter and energy can pass through the boundaries.

● Is it possible to add energy in other ways?

A system that is enclosed by a wall which does not permit the passage of matter or energy is called an *isolated system*; one that allows energy but not matter to cross the boundary is a *closed system*; one that allows the entry of both matter and energy is an *open system*. These possible situations are depicted in Fig. 2-1.

Some specialized closed systems do not permit the passage of heat. These are very basic in thermodynamic theory and are called *adiabatic* systems. Diathermal walls, on the other hand, allow the passage of heat.

2-2. Examples

The *earth* is a closed system because it does not receive or expel significant amounts of matter (the amounts lost into space or gained through meteorite falls, etc., are negligible). It is not an adiabatic system, as heat is both received from and reradiated into outer space (see Fig. 2-2a).

A *cell* (Fig. 2-2b) is an open system. Both matter and energy can go through its boundary, the cell membrane. We should keep in mind, however, that the cell membrane is differentially permeable, so some substances may not enter or leave the system.

A common *coffee cup* with its cover is, for practical purposes, an adiabatic system. Strictly, it is not a real adiabatic system because it eventually loses its heat (Fig. 2-2c).

Biological systems obey the same general rules that apply to inanimate systems. The complexity of most biological energy exchanges, however, makes it mandatory to be extremely careful in setting up systems descriptions. This is illustrated in Fig. 2-3, which shows an example of some of the energy inputs and outputs in a whole animal.*

*One of the most dangerous pitfalls is the assumption that perceived quantities necessarily correspond to physical reality. A striking example is the perception of temperature; it has been stated that the concept of temperature measurement is based on the physiological sensations of hot and cold. Actually, the human body is a poor thermometer; the presence of a layer of stagnant air, the' boundary layer, insulates the body from the environment and prevents us from perceiving the actual temperature of the surroundings. If this layer is removed, using convection currents generated by a fan, for example, heat exchanges with the environment occur and the external temperature is perceived. This is the reason the air coming from a fan feels colder than the surrounding stagnant air.

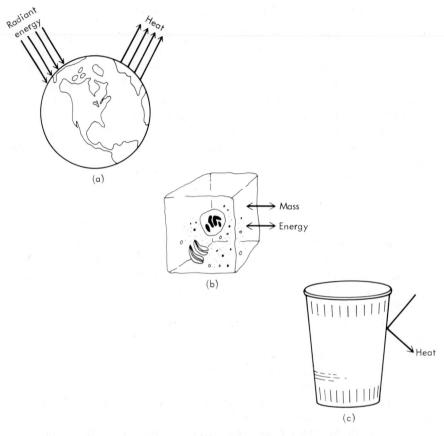

Fig. 2-2. Examples of thermodynamic systems. (a) The earth is a closed system, as mass, for all practical purposes, is neither gained nor lost. Notice that heat can enter and leave, so that the system *is not* adiabatic. (b) A cell is an open system; both matter and energy go through the boundaries. (c) For all practical purposes, a coffee cup is an adiabatic system, as heat cannot leave the container.

2-3. Equilibrium

The popular notion that an object is in equilibrium when it does not move is a perfectly valid one. In thermodynamics we extend this concept to encompass systems in equilibrium; we say that a system is at an equilibrium state when none of its properties changes in time.

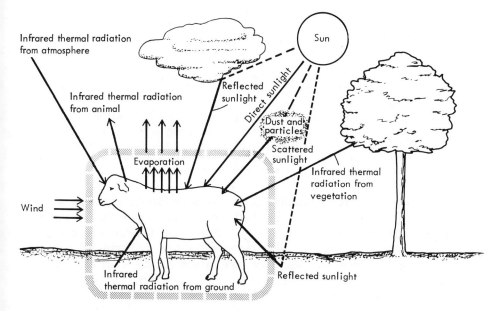

Fig. 2-3. The exchange of energy between a living organism and its environment can also be described using the system approach. The complexity of even the most superficial biological energy exchanges, however, makes it mandatory to exercise extreme care in setting up a complete description. (Adapted from D. M. Gates, "The Energy Environment in Which We Live," *American Scientist*, vol. 51, p. 327.)

2-4. Intensive vs. Extensive Variables

An isolated system will obviously reach an equilibrium state that is completely independent of the surroundings, but an open system will interact completely with the surroundings and change them as well, as shown in Fig. 2-4. A closed system is somewhere in between these extremes: If the external variables (surroundings) do not change in time, they uniquely determine the values of all the internal variables at the equilibrium point. What are these variables? We can distinguish two general types

- *Intensive variables*, which at equilibrium are independent of the volume or size of the system

- *Extensive variables*, which are a function of the size of the system under consideration

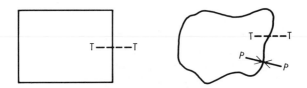

Fig. 2-4. Once the external intensive variables are determined, the equilibrium state of a closed system is completely specified. In a diathermal system with fixed boundaries, for example, the external temperature determines the equilibrium state; if the walls can change shape, both the pressure and temperature must be specified.

Temperature, for example, is an intensive variable, because any part of a system which has reached equilibrium will have the same temperature; furthermore, a system whose walls permit the passage of heat will have the same temperature as the environment when equilibrium is attained. *Mass,* on the other hand, is an extensive variable, because the amount of mass will depend on what portion of the volume we analyze.

2-5. Not All Variables Are Needed to Specify the Equilibrium States Uniquely

Measurable quantities such as pressure, volume, temperature, and composition are not completely independent of one another at equilibrium; they are instead connected by a unifying formula called *equation of state,* which relates the variables to one another. In a gas, for example, if the pressure, volume, and number of molecules are specified, the temperature is automatically determined. The general equation of state may be hard or easy to find depending on the complexity of the system. One of the simplest systems—which we shall have opportunity to consider later—is an "ideal" gas, which obeys, at equilibrium, the equation of state

$$PV - nRT = 0 \qquad \text{or}$$

(2-1) $$f(P, V, n, T) = 0,$$

in which P, V, n, and T are the pressure, volume, number of moles, and temperature, respectively.* That is, there is a function of the

---

*A mole of any substance contains 6×10^{23} molecules of that substance.

pressure, volume, number of moles, and temperature which is identically zero at any equilibrium state. In general, there will be a function of all the measurable variables of the given system such that the equation is identically zero at equilibrium.

2-6. Functions of State

All the variables that are uniquely defined at an equilibrium state are called *functions of state* (not to be confused with equations of state defined above). The central problem in thermodynamics is to find out which variables are functions of state and, once they have been found, to calculate their changes between two equilibrium points. Thermodynamics, however, does not consider the *absolute* values of these quantities, only their changes. The calculation of values *at* a given state has to be done by assuming specific microscopic models of matter. Since nothing can be stated about the intermediate values of these quantities, the "time" variable is foreign to classical thermodynamics: Equilibrium thermodynamics, then, should be more properly referred to as "thermostatics." Moreover, the only processes that can be precisely described by thermostatics pass through intermediate states which are, like the initial and final states, equilibrium states. These processes are called *reversible* processes. Clearly, it is impossible in practice to bring about a transition or process if the system is in equilibrium—if it does not move—at each instant of time.

Leonard K. Nash of Harvard University summarizes the situation in the following terms: "[Classical] Thermodynamics is, then, if you please, rather more the science of the possible in principle than the science of the attainable in practice."

2-7. Is There Anything Special About Functions of State?

Since changes in functions of state are not dependent on intermediate states but only on initial and final values, *their change between any two arbitrary equilibrium states is independent of how the process takes place.* This single property makes them very useful because if we can find the change in a given function of state for *any process*, we are assured that the change will be the same for any other process the system undergoes, provided that the initial and final states are those considered originally.

2-8. Cyclic Processes

Most engines (biological or inanimate) are designed to perform a given job in a periodic fashion; that is, they repeat a few steps over and over. The heart is an example of such a machine. In thermodynamic terms we would say that the machine starts at a state S_1, goes through a series of states, and returns to the original state S_1. Suppose that a machine works in such a cyclic manner and assume, to make things very simple, that there are only two equilibrium states, 1 and 2, as shown in Fig. 2-5, which are uniquely specified in terms of all the intensive and extensive variables at these states.

Consider now the change in some function of state E when the system goes from state 1 to state 2. The change in the value of the function of state can be found from the differences between the final and initial values,

$$(2\text{-}2) \qquad\qquad \Delta E_{1 \to 2} = E_2 - E_1,$$

in which the Greek letter delta is used as a shorthand notation to represent the difference between the two values. The change in the same quantity E when the system goes from 2 to 1 is now

$$(2\text{-}3) \qquad\qquad \Delta E_{2 \to 1} = E_1 - E_2.$$

The sum of the two changes will give the change in this function when the system goes around the full cycle. Clearly, the sum of the two changes is zero, which means that the function of state is

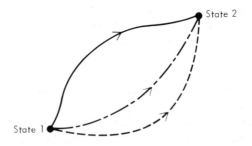

State 2

State 1

Fig. 2-5. Functions of state are calculated between two equilibrium states, which can be specified in terms of measurable quantities such as temperature and pressure. The change in a function of state is the same for any path going between the same initial and final states and is also independent of the process—any system that goes from state 1 to state 2 will show the same change in a function of state regardless of its physical design.

conserved in a cyclic process; that is, does not change when the system goes around a cycle.

2-9. Some Functions of State Are Defined in Terms of Combinations of Other Quantities Which Are Not Themselves Functions of State

The simplest example of a function of state is provided by a variable that can be measured at equilibrium, such as temperature, volume, and concentration. Unfortunately, these quantities are the least useful from the thermodynamic viewpoint.

Those functions of state which have thermodynamic value are usually hidden; nature does not reveal them immediately and we must discover them; to make life even harder, we cannot in general make an instrument and measure their value. How can we find a function of state which we cannot measure directly? The method consists in taking a system between two states following different paths and showing that there is a *combination of measurable quantities* which always gives the same change no matter what path is taken. This new quantity will then be a function of state. We can infer that it has uniquely defined values at the end equilibrium points even though we may not be able to tell what meaning the new function has in the real world. Once we have found a function of state, we automatically know that it is conserved in a closed cycle.

2-10. Energy Is a Function of State

The First Law of Thermodynamics states that there exists a quantity, the internal energy of a system, which is a function of state in a closed system in equilibrium. Although the First Law does not say *how* to find the value of the internal energy *at* a given equilibrium state, it gives a recipe for finding *changes* in energy between two equilibrium states: The internal energy change between two states equals the heat added to the system from the environment minus the work done by the system on the environment (Fig. 2-6); that is,

$$(2\text{-}4) \qquad\qquad \Delta E = Q - W,$$

in which Q is the heat added to and W the work done by the system. As we shall see, neither Q nor W are functions of state. In order to find that a function E existed, it was necessary to recognize that heat

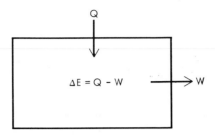

Fig. 2-6. The First Law of Thermodynamics states that energy is a function of state whose change between two states is calculated by subtracting the work done by the system on the environment from the heat added by the environment to the system.

and work were forms of energy, which was no small intellectual achievement.

An understanding of the meaning and application of the First Law of Thermodynamics requires some familiarity with the concepts of work and heat. We all have an intuitive idea of what work and heat are. This is very unfortunate because intuitive notions rarely fit physical reality; while we tend to equate heat with temperature and work with subjective effort, physical heat can be extracted from a cold region and an individual may exhaust himself while performing no physical work. A short digression on work and heat is then in order.

2-11. Work

Physical work is only vaguely related to the intuitive concept of work. Whenever *motion* arises as a result of the application of a force on an object, we say that the work done by the force is given by the product of the force times the distance the object is moved:

(2-5) $$W = F \times d.$$

This definition holds *only* if the force remains constant during the motion *and* if the force is applied in the direction of motion. If the force is perpendicular to the direction of motion as shown, for example, in Fig. 2-7(a) no physical work is done. Similarly, if no motion results when the force is applied, no work is done either (since the distance moved is zero). Thus, we may get tired by holding a heavy weight in the air but, rigorously speaking, we are not doing any work on the weight as long as we do not move it.

Work is done either by the system on the environment or by the

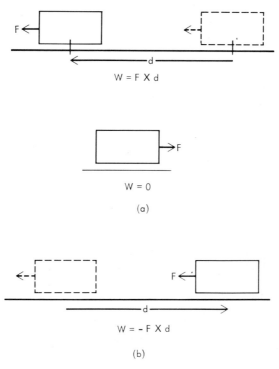

Fig. 2-7. (*a*) Mechanical work is given by a force times the distance it displaces. If a force is present but the object does not move, no work is done. (*b*) If the object displaces *against* the direction of the force, the work done is negative.

environment on the system. If the force is opposite to the direction of motion, as shown in Fig. 2-7(*b*), the work is negative and the object is doing work on the force rather than the force on the object; that is, the system does work on the environment.

2-12. How Work Is Calculated in Specific Cases

Although we give the analytical approach to the calculation of work using calculus in the Appendix, it is convenient to have a general idea of how to find the work done by an engine when there is some information as to its behavior. Most of the time the work done cannot be calculated using formulas; there is, however, a convenient graphical method devised by James Watt, the inventor of the first successful steam engine.

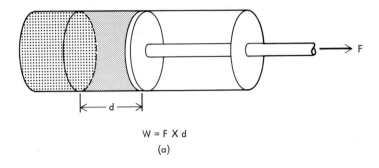

$$W = F \times d$$

(a)

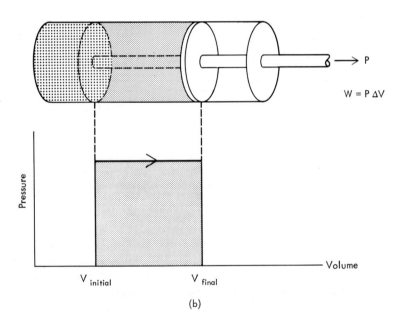

$$W = P\,\Delta V$$

(b)

Fig. 2-8. (a) The work done by an expanding gas is calculated by multiplying the force on the piston times the displacement, but it can also be expressed as (b) the gas pressure times the volume change. This corresponds to the shaded rectangular area in the pressure-volume diagram.

Watt considered the expansion of a gas inside a piston, as shown in Fig. 2-8(a). In this simple case heat is drawn from a bath held at constant temperature, and work is done by the gas on the environment. We have seen that if the force is constant, the work will be given by Eq. 2-5. It is more convenient when considering gases to give the work in terms of the volume change and the internal pressure of the gas. The pressure is the force applied to a unit area and it is defined as

(2-6)
$$P = \frac{F}{A},$$

while the volume change is the displacement, d, times the cross-sectional area of the piston, shown in Fig. 2-8(b), top.

(2-7)
$$\Delta V = A \times d.$$

The work can then be rewritten as

(2-8)
$$W = \frac{F}{A} \times d \times A,$$

in which we have both multiplied and divided Eq. 2-5 by the area of the piston. Since F/A is the pressure of the gas, and $d \times A$ is the volume change ΔV, the work is also given by

(2-9)
$$W = P \times \Delta V.$$

This equation will be correct provided the pressure, hence the force, is constant. If we plot the external pressure as a function of the volume, the expansion of the gas is represented by a horizontal line as shown in Fig. 2-8(b); the work term, $P \Delta V$, is then the rectangular area shaded in the figure. We have added an arrow on the constant pressure line to indicate the direction of the transition. Notice that when the arrow points from left to right the system expands and does work on the environment; if the arrow points in the opposite direction the environment does work on the system, and the gas compresses.

What happens if the pressure changes with the displacement? This could easily be done, for example, by controlling the motion of the piston by hand. In this case we could subdivide the process into several steps each having a force that is about constant during the step. For example, if the pressure changes three times as shown in Fig. 2-9, the work will be given by the sum

(2-10)
$$W = P_1 \Delta V_1 + P_2 \Delta V_2 + P_3 \Delta V_3$$

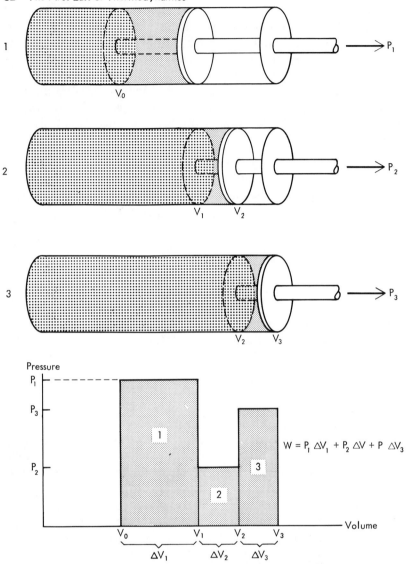

Fig. 2-9. If the pressure is not constant but changes during the expansion, the work is calculated by breaking the process down into several steps, such that the pressure is constant (or almost constant) during each step. The total work done can then be found by adding up the areas of the individual rectangles.

which, in a pressure volume diagram, would be the area of the three rectangles shown in the drawing. As before, the arrow points from left to right so the system does work on the environment.

In general, if the pressure changes in some arbitrary manner, we can still subdivide the graph into small rectangles such that the force will always appear to be constant for small displacements. The total work will then be the area under the pressure-volume curve.

Consider now a diagram corresponding to an engine which works on a cycle so that it always returns to its original state, as shown in Fig. 2-10. Since part of the work is done by the system on the environment (upper portion) while part of it is done by the environment on the system (lower part, arrow goes from right to left) the net work is the difference of the two areas—which is the area enclosed by the original cycle. It is important to stress that in order to calculate work this way one must assure that all the points in the pressure-volume curve are actually traversed. This will happen only if the process takes place very slowly and follows equilibrium points.

2-13. Work Can Be Converted into Kinetic Energy

The earliest concept of energy was given in terms of the energy carried by a moving body,

$$(2\text{-}11) \qquad\qquad \text{K.E.} = \tfrac{1}{2} M v^2 ,$$

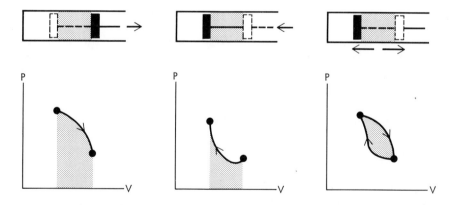

Fig. 2-10. In general, if the pressure changes during an expansion, the work is given by the area under the *P, V* curve. If the arrow goes from left to right (expansion), the system does work on the environment. If the arrow goes from right to left (compression), the environment does work on the system. During a cycle in the clockwise direction, the total work done is the expansion work minus the compression work. This is the shaded area enclosed by the curve.

in which M is the mass of the object and v its velocity. This quantity was called *vis viva*, or living force, during the last century. We now refer to it as *kinetic energy*.

When the object is allowed to fall from a height H starting from rest, it is found that the higher the initial height the larger the final velocity before it strikes the ground. Moreover, if the object is thrown upward, the larger its initial velocity the higher the height reached. It appeared, then, that the energy of motion could be stored as another form of energy, potential energy, which was somehow related to the height itself. If the potential energy is written as

(2-12) $$\text{P.E.} = MgH$$

—in which g is the acceleration of gravity, M the mass of the moving body, and H the height—the quantity defined by the sum

$$E = \text{P.E.} + \text{K.E.}$$

is constant at any height. At a given height, the potential energy may be larger than the kinetic or the kinetic energy larger than the potential. Their sum is, however, a constant. This result could also be stated "energy is conserved."

What is the mysterious potential energy term MgH? As it turns out, it is simply the work done by the gravitational force. According to the second law of Newton, the force on an object is measured by the acceleration the force imparts to the object,

(2-13) $$F = Ma,$$

and since the acceleration of gravity on earth, g, is the same for any object, the term Mg is clearly the weight of an object. The product MgH is then the work done to displace this weight a distance H.

The question now arises as to what happens when the object hits the ground. The kinetic energy is zero and the potential energy is also zero; our conservation formula then stops from being conserved at that time. We may inquire as to what has been the effect of the collision on the body and the surface. If the moving object falls on a very soft surface, like talcum or sand, it is found that the surface has deformed; if, on the other hand, it falls on a hard surface and neither the object nor the surface are distorted as a result of the collision, both the object and the surface heat up. We now say that energy has transformed into work or heat, respectively. While the transformation of kinetic energy into mechanical work of deformation was perfectly clear, the conversion of energy into heat was not.

2-14. Is Heat a Fluid?

It may seem amazing to us that the casual connection between energy and heat was not recognized, but the explanation given at the time also made perfect sense given the knowledge then available. The most widely accepted theory of heat claimed that it was a fluid contained in all bodies and that it could be extracted by friction. When two objects collided, the appearance of heat could easily be accounted for in terms of the release of "fluid" caused by friction.

This fluid was designated by the name caloric and was supposed to have fancy properties like weight and self-repulsion. Count Rumford—Benjamin Thompson, born in the United States—disproved the caloric theory in a number of interesting experiments designed to show that heat could not be a material agency. His most popular experiment is probably the observations on cannon boring published under the title "An Inquiry Concerning the Source of Heat which Is Excited by Friction." In these experiments Rumford utilized a setup to bore cannons and showed that it is possible to extract as much heat as one wishes by applying continuous friction to a piece of metal. The caloric theory, then, was inconsistent: If heat were a fluid, a given body would contain only a finite, not infinite, amount of it. After this and other experiments (Fig. 2-11) in which he showed that heat did not have weight and that it could be transferred through vacuum (both in contrast with the caloric theory), Rumford correctly proposed that heat was a form of motion, or energy, but his ideas were not readily accepted.

2-15. Metabolic Heat and Energy Conservation

Credit for the recognition that heat is a form of energy usually goes to Julius Mayer, a physician who on a sailing trip in the Carribean, according to legend, noticed that the venous blood of sailors in warm climates had a distinctive bright red color, a sign of the presence of large amounts of oxygen in the blood. Venous blood has given up most of its oxygen to tissues, under normal conditions, so it is usually dark brown. Mayer thought that if heat were a form of energy, this observation would make perfect sense: The body would require less fuel—measured in terms of the oxygen used to burn the "fuel"—in warm regions to maintain a constant temperature because energy can be acquired directly from the environment in the form of heat. Although the explanation is basically correct, it took some

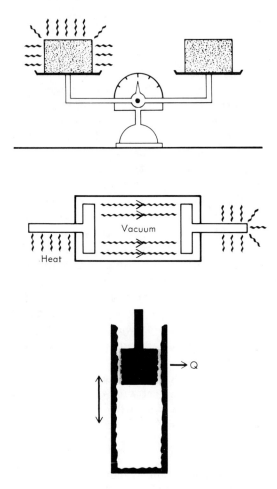

Fig. 2-11. Some observations which indicate that heat does not have the properties of a fluid (matter): A hot block and a cold block of the same material and the same size weigh the same; heat is transmitted through vacuum; an indefinite amount of heat can be extracted from a solid by friction. These observations were originally made by Count Rumford.

careful measurements by James Prescott Joule to show the equivalence between heat and other forms of energy.

2-16. The Mechanical Equivalent of Heat

Joule reasoned that if Mayer was correct and heat was a form of energy, different kinds of energy, such as electrical or mechanical, should release the same amount of heat when completely dissipated inside a calorimeter.

As depicted rather freely in Fig. 2-12, Joule performed various experiments using mechanical and electrical energy which showed that this was the case. Furthermore, he measured the conversion factor between heat and mechanical energy and found that 1 calorie—the heat required to change the temperature of 1 gram of water by 1 degree centigrade—was equivalent to 41,850,000 ergs, an erg being the mechanical energy required to accelerate 1 gram of matter 1 centimeter per second squared in 1 centimeter. A comparison with other units of energy is given in Fig. 2-13.

2-17. Energy Is a Function of State

The experiments of Joule showed that the energy of a closed system could be changed by adding heat or any other form of energy. Since all forms of energy can be converted into heat, it was clear that in order to change the temperature inside the calorimeter by a certain

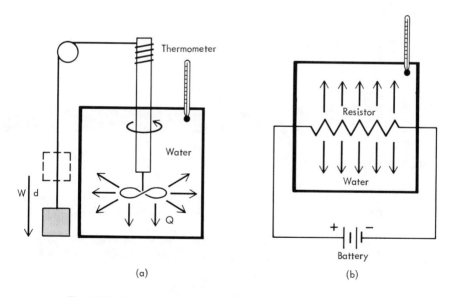

Fig. 2-12. By transforming various forms of energy into heat inside a calorimeter (an adiabatic container), Joule showed that the same amount of heat appeared in the system when the same amount of any form of energy was dissipated. Thus, if the mechanical and electrical work done in (a) and (b) is the same, the temperature changes in both calorimeters will be equal.

Fig. 2-13. Equivalence between various energy units. Most energy units originate through the formula for work, $W = F \times d$ or, since $F = m \times a$, $W = m \times v/t \times d$, at constant acceleration. When the basic units chosen for mass, distance, and time are grams, centimeters, and seconds, the work is given in "ergs"; if the basic units are kilograms, meters and seconds, the resultant energy is given in units of "joules". Other units used are the calorie (the most common in biology) and electron-volt (used by physicists). The watt measures rate of energy utilization or generation and equals one joule per second. (Reproduced from C. Starr, *Scientific American*, September 1971, vol. 224, p. 49.)

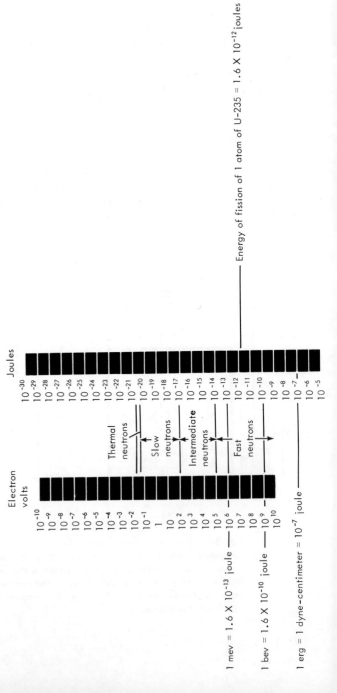

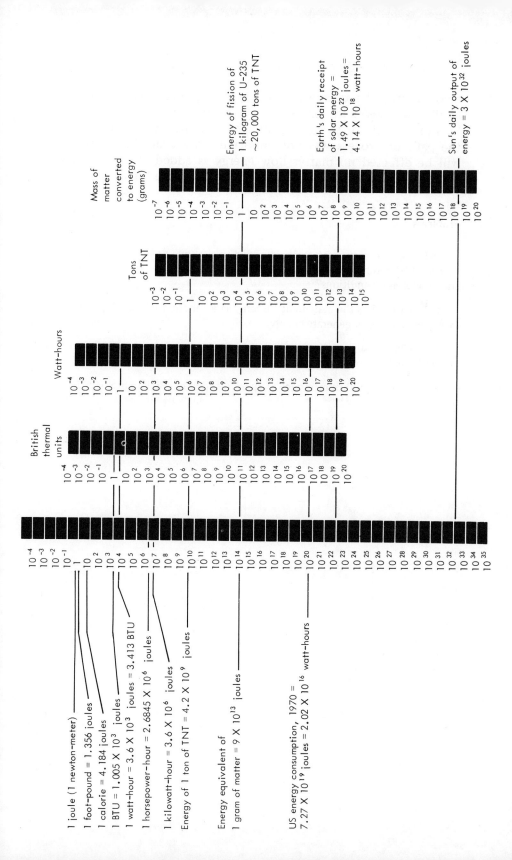

amount—in order to move from a given equilibrium state to another equilibrium state—the process could be performed by adding heat or by doing work, or by any combination of the two. In other words, the path or process was not important; the same change in energy will be effected no matter how energy is added. This implies that energy is a function of state; hence it is conserved in a closed cycle.

Problems

2-1. Can we ever write the heat added to a system as $Q_{final} - Q_{initial}$; that is, can Q be a function of state under any conditions? Can W be a function of state under any conditions?

2-2. Show that the statement "Energy is a function of state" implies that a cyclic engine cannot generate energy.

2-3. What assumptions—in terms of efficiency of energy conversion—are made in the measurement of the mechanical equivalent of heat?

2-4. Is the equilibrium model of thermostatics relevant in biology? Discuss.

3

THE DIRECTION
OF NATURAL PROCESSES:
ENTROPY AND FREE ENERGY

As a student, I read with advantage a small book by F. Wald
entitled "The Mistress of the World and Her Shadow." These
meant Energy and Entropy. In the course of advancing knowledge
the two seem to have exchanged places. In the huge manufactory
of natural processes, the principle of entropy occupies the position
of manager, for it dictates the manner and method of the whole
business, while the principle of energy merely does the bookkeeping,
balancing credits and debits.—Robert Emden, 1933 *Nature,*
vol. 141, p. 908.

Energy is useful only when it can be converted into work. The First
Law of Thermodynamics gives a definite relationship between the
energy input, the work done by the system, and the energy lost in
the form of heat, but it does not provide information as to how
much of the energy added will actually be used to do work.

As it turns out, there are serious restrictions on how efficiently
energy transforms into work. These are given by the Second Law of
Thermodynamics. In essence, the Second Law states that there is
always a certain amount of energy which changes into a "lower
quality" form such as heat and becomes less available to do work.
This is not caused by an inherent problem of design in any particular
engine or process; it is a law of the universe which applies, as far as
we know, from the smallest atom to the largest galaxy.

3-1. The Second Law of Thermodynamics

The original statements of the Second Law of Thermodynamics appear to be completely unrelated to one another:

- Heat cannot pass spontaneously from a low to a high temperature level (Rudolf Clausius).
- It is impossible to continuously produce work by means of inanimate material agency cooling only one body down to a temperature below the coldest part of the surroundings (Lord Kelvin).
- It is impossible in a system enclosed by an envelope which permits neither a volume change nor passage of heat, and in which both the temperature and the pressure are everywhere the same, to produce any inequality of temperature or pressure without the expenditure of work (James Clerk Maxwell).
- It is impossible to design a machine that completely transforms heat into work (Sadi Carnot).

These rather negative formulations of the Second Law are depicted in Fig. 3-1, but their enigmatic forms can be clarified only after considering the mathematical form of the law. The mathematical expression of the Second Law is introduced through a new function of state, the entropy S, which always increases in spontaneous isolated processes and which also has the quasi-mystical attribute of pointing the direction of time.

3-2. Reversible vs. Irreversible Processes

Fundamental in the development of the entropy concept is the distinction between spontaneous, or irreversible, and reversible processes. It is a simple matter to distinguish irreversible from reversible transitions: If a reversible process is filmed, it will be possible to project the resultant movie reversed in time—from the end toward the beginning—without giving away the fact that the movie has been reversed. A movie of a spontaneous process, on the other hand, looks absolutely ridiculous when it is reversed: People walk backward; smashed cars reassemble spontaneously and start running away from each other (and backward, of course); a swimmer jumps from the bottom of a pool 9 feet up in the air and so on . . .

It is for this same reason that the processes described in the

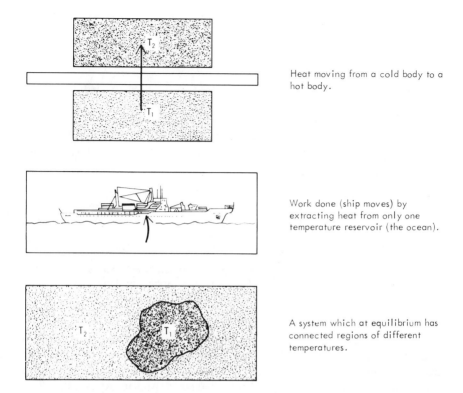

Heat moving from a cold body to a hot body.

Work done (ship moves) by extracting heat from only one temperature reservoir (the ocean).

A system which at equilibrium has connected regions of different temperatures.

Fig. 3-1. Some spontaneous processes that the Second Law says are impossible.

statements of the Second Law given in Sec. 3-1 are impossible: They all consist of time-reversing processes which are irreversible. We can then conclude that the Second Law establishes that there is only one possible direction of time for irreversible processes. This is the one that increases a function of state called *entropy*, which prompted the physical astronomer Sir Arthur Eddington to say that entropy points the arrow of time (Fig. 3-2): All irreversible processes can go only forward in time while reversible processes can go forward or backward. By performing the "movie reversal" test on various processes, we quickly conclude that most processes are irreversible.

3-3. Entropy

Heat and work, Q and W, are not functions of state, since their values are determined by the way in which a given process takes place. It is

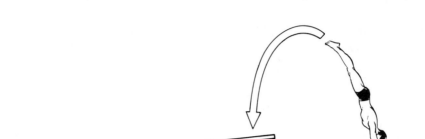

Fig. 3-2. The movie of an irreversible process projected backward in time looks intuitively wrong. Physically, the wrong direction of time decreases entropy.

sometimes possible, however, to associate a function of state with a quantity that is not itself a function of state. The mathematical trickery involved does not add too much to the physical understanding and it is therefore discussed in the Appendix. The main result which is of interest to us is that it is possible to find a function of state which is associated with reversible heat transfers.

It is found that the ratio of the heat gained during a reversible process at constant temperature T divided by the temperature,*

$$\frac{Q_{\text{rev}}}{T}$$

is independent of how the process occurs, as long as the various transitions take place between the same initial and final states (Fig. 3-3). This is a remarkable result because it says that the ratio given above is the change in a function of state, even though Q_{rev} by itself is not. We can denote the change in this quantity by

(3-1) $\Delta S = S_{\text{final}} - S_{\text{initial}} = \frac{Q_{\text{rev}}}{T}.$

---

*Thermodynamic temperatures are given in "absolute" degrees. The conversion formula between absolute (Kelvin) degrees and degrees centigrade is

Temperature (Kelvin) = temperature (centigrade) + 273.

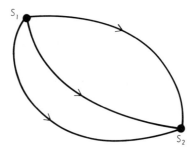

Fig. 3-3. Although entropy can be calculated only for a *reversible* process between two equilibrium paths, all other processes (including irreversible) that go between the same initial and final states will have the same change in entropy.

The new function of state S is called the *entropy* of the system. Although the study of the origin and hidden meaning of words is best left to archaic English courses, it is interesting to point out that entropy comes from a Greek word meaning evolution. This is precisely what entropy does: It points the direction in which natural processes will evolve.

If heat is lost rather than gained, the *reversible* change in entropy is given by

$$\Delta S = \frac{-Q_{rev}}{T}.$$

Therefore, when the system gains heat reversibly from a bath held at constant temperature T (Fig. 3-4), the *total* change in entropy for the system plus surroundings (bath) is

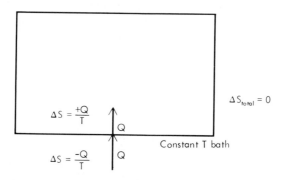

Fig. 3-4. The total entropy of the system plus the environment does not change during a reversible heat transfer at constant temperature.

$$\Delta S = \Delta S_{\text{system}} + \Delta S_{\text{bath}}$$

$$= \frac{Q_{\text{rev}}}{T} + \frac{-Q_{\text{rev}}}{T} = 0.$$

Thus, there is no net change in entropy during a reversible process.

3-4. Entropy Changes During Irreversible Processes

The mathematical expression of the Second Law states the following:

- During an irreversible isolated process the entropy of the system plus its environment *increases*—it is not zero.
- At equilibrium entropy reaches its maximum value.

Consider again the transfer of heat from a bath at constant temperature to a system in which the absorption of heat is irreversibly done. Let us say, to make life simple, that the removal of heat from the bath is *reversible*. The change in entropy for the system plus environment is

$$\Delta S_{\text{total}} = \Delta S_{\text{reservoir}} + \Delta S_{\text{system}}$$

and we know that $\Delta S_{\text{reservoir}} = \dfrac{-Q}{T}$

because that part of the process is reversible. If the absorption of heat by the system were also reversible, the change of entropy for the system would equal $+Q/T$ and would exactly balance the entropy decrease of the bath. The Second Law says, however, that the entropy change *is not* zero; it is larger. It then follows that during the irreversible change the entropy increase of the system must be *larger* than Q/T. We can write the expression either as an inequality,

$$\Delta S_{\text{irrev}} > \frac{Q}{T}$$

or as an equality

(3-2) $$\Delta S_{\text{irrev}} = \frac{Q}{T} + \Delta_i S,$$

in which $\Delta_i S$ is the *internal entropy production*, the extra amount of entropy introduced by the irreversible process. Clearly, $\Delta_i S = 0$ for a reversible process.

3-5. Entropy Is a Function of State!

Even though for a given *amount of heat added* the change in entropy is smaller for a reversible than for an irreversible transition, the change in entropy for an arbitrary change *between the same initial and final states* is independent of whether the process is reversible or irreversible. This has to be true because entropy is, like energy, a function of state. Of course, the two processes *will not* be the same in the sense that different amounts of heat will be added in the reversible and irreversible processes. Otherwise, we would be back to the previous case in which

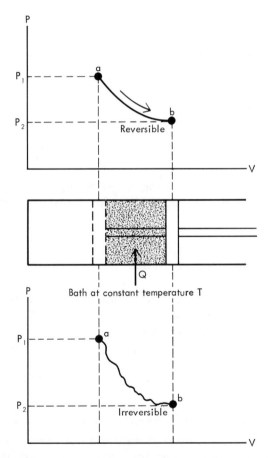

Fig. 3-5. The change in entropy *for the system* is the same when the gas expands reversibly or irreversibly. The *total* change in entropy for system and surrounding, however, is larger in the irreversible process.

$$\Delta S_{\text{irrev}} > \Delta S_{\text{rev}}$$

and then the transition would not proceed between the same *states* (Fig. 3-5).

Consider two such *processes* taking place *between the same initial and final states*. Let the first be reversible and the second irreversible. The change in entropy for the first process is

$$\Delta S_1 = \frac{Q_1}{T_1},$$

while the change in entropy for the second, irreversible process is

$$\Delta S_2 > \frac{Q_2}{T_2}.$$

According to the previous discussion,

$$\Delta S_1 = \Delta S_2.$$

it then follows that

$$\frac{Q_1}{T_1} > \frac{Q_2}{T_2}$$

and if both processes take place at the same temperature,

$$Q_1 > Q_2;$$

that is, the heat gained during a reversible isothermal transition between two states is larger than the heat gained during an irreversible transition.

3-6. Entropy Changes and Work

What is the energy gained by the system in the isothermal transfer of heat from the bath? According to the First Law, the energy change during the reversible process is given by

$$(3\text{-}3) \qquad E_{\text{final}} - E_{\text{initial}} = \Delta E_{\text{rev}} = Q_{\text{rev}} - W_{\text{rev}}$$

while the expression for the energy change during the irreversible process is

$$(3\text{-}4) \qquad E_{\text{final}} - E_{\text{initial}} = \Delta E_{\text{irrev}} = Q_{\text{irrev}} - W_{\text{irrev}}.$$

But since energy is a function of state and both processes go between the same final and initial states, the two changes must be equal

$$Q_{\text{rev}} - W_{\text{rev}} = Q_{\text{irrev}} - W_{\text{irrev}}.$$

After rearrangement, this expression gives

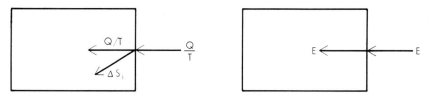

Fig. 3-6. Entropy and energy behave differently when crossing the boundary of a system. Although the same energy appears inside the system as crosses the boundary, an excess entropy is "generated" inside the system in irreversible processes; we denote this additional amount of entropy by ΔS_i.

$$(3\text{-}5) \qquad\qquad Q_{rev} - Q = W_{rev} - W_{irrev}$$

and since the heat gained reversibly is larger than Q_{irrev} we conclude that *the work done by the system during the reversible process is larger than the work done by the irreversible process.* This is a very important conclusion of the Second Law of Thermodynamics. Figure 3-6 contrasts the behavior of energy with that of entropy; although both are functions of state, energy is conserved for the system plus the environment while entropy always increases.

3-7. The First and Second Laws Combined

We previously wrote the First Law for a closed system in its most general form,

$$\Delta E = Q - W.$$

Since the heat added reversibly to the system is related to the entropy change by

$$\Delta S = \frac{Q_{rev}}{T},$$

we obtain
$$\Delta E = T\,\Delta S - W_{rev}.$$

This formula combines the first and second laws. In the particular case in which pressure-volume work is done, we can write

$$(3\text{-}6) \qquad\qquad \Delta E = T\,\Delta S - P\,\Delta V.$$

3-8. Free Energy: The Biological Function of State

Although the Second Law of Thermodynamics has universal validity, the actual calculation of entropy changes in specific cases can be elaborate. This is a result of the wide generality of the entropy principle. As a result it is sometimes convenient to find other functions of state which can act as indicators of equilibrium and irreversibility and which are related to real measurable quantities. There are an infinite number of functions of state which can be generated from the internal energy by adding and subtracting variables, their choice must be dictated by the types of constraints imposed on the system. Thus, in actual experiments not all variables are changed at once when a transition between two states takes place; we may change the temperature keeping the composition constant or change the composition and keep the pressure constant, etc. In particular, biological transformations always take place at constant temperature and, very often, at constant pressure. This is also true of chemical reactions which take place in an open beaker: The temperature of the reaction is the same as that of the room where the beaker is, while the pressure is atmospheric pressure.

Under conditions of constant temperature and pressure, it is convenient to introduce a new function of state, the *Gibbs free energy* of the system defined by the equation

$$(3\text{-}7) \qquad G = E + PV - TS.$$

The introduction of G and its definition are, we should stress, completely arbitrary. As it turns out, however, at constant temperature and pressure, the Gibbs free energy gives a direct measure of the maximum amount of work a system can perform. It gives, at the same time, a convenient method for testing biological equilibrium in a quantitative fashion.

Some important properties about the free energy G which are proved in problems at the end of the chapter are:

- Free energy *decreases* during a spontaneous (irreversible) process at constant temperature and pressure.
- The maximum amount of work that can be obtained from a process taking place at constant temperature and pressure is equal to the free-energy decrease of the system. This is true only in reversible processes; for irreversible processes the work done is less than this amount.
- The free-energy change at constant temperature and pressure is given by

$$(3\text{-}5) \qquad \Delta G_{TP} = - T \Delta_i S - W',$$

in which $\Delta_i S$ is the internal entropy production introduced in Sec.

3-4 and W' the "useful" work—defined as all the work that is not pressure-volume work.

3-9. The Useful Work and ΔG

In order to derive an explicit expression for the free energy change ΔG, we must first give a formula for the "useful" work W'. We should stress that there is nothing particularly useful or useless about the "useful" work. It is just *defined* as all the work that is not pressure-volume work; that is,

(3-8) $$W' = W - P\,\Delta V.$$

But the question still remains: What is the most general expression for the work done? We have already introduced the reversible pressure-volume work given by

$$W = P\,\Delta V.$$

and the force-distance work, $F \times d$.

Two other forms of work of interest in biological systems are the work required to bring mass into a region where mass is already present and the work required to bring a charge into a region where charges are already present (Fig. 3-7). In general, both of these forms of work depend on the particular geometry considered, composition of the system, and other factors.

We can give a general expression for the work of transport of mass and charge by analogy with the pressure-volume expression. The calculation of pressure-volume work is done by multiplying an intensive property, the internal pressure, by the change in an extensive variable, the volume. By analogy, during the performance of electrical work there is an extensive variable, the charge q, which enters or leaves the system. We can *arbitrarily define* a new intensive property, the electrical potential ψ, which must be found for the given system. The electrical work is then

(3-9) $$W_{elec} = -\psi\,\Delta q$$

The negative sign appears because when the system does work on the environment, the positive charge in the system *decreases*. Similarly, we can *define* the work of mass transport as

(3-10) $$W_{mass} = -\mu\,\Delta n,$$

in which μ is an intensive variable called the *chemical potential* and

Fig. 3-7. Pictorial representation of the basic types of work done by biosystems: electrical work, mass transport, chemical reactions.

Δn the amount of mass that enters the system. Again, the system does work on the environment when mass *leaves*, thus the need for the negative sign.

3-10. The Explicit Expression for the Free-Energy Change

In order to put free energy to use in practical cases, we must have some way of relating the free-energy change to measurable quantities. As in the case of entropy, we merely need to find the free-energy change for a reversible process, since free energy is a function of state and any other process *between the same two states* will show the same increase or decrease in free energy. The first step consists in writing the total expression for the First Law of Thermodynamics when all the work terms are given. The resultant change in energy is

$$\Delta E = Q_{rev} - W_{rev}$$
$$= T\,\Delta S - [+P\,\Delta V - \mu\,\Delta n - \Psi\,\Delta q - f\,\Delta L]$$
$$= T\,\Delta S - P\,\Delta V + \mu\,\Delta n + \Psi\,\Delta q + f\,\Delta L.$$

Introducing this formula into Eq. 3-4, we obtain an expression for the free-energy change during a reversible process *at constant temperature and pressure*

$$\Delta G_{T,P} = \Delta E + P\,\Delta V - T\,\Delta S$$
$$= [T\,\Delta S - P\,\Delta V + \mu\,\Delta n + \Psi\,\Delta q + f\,\Delta L] + P\,\Delta V - T\,\Delta S$$

which, after cancellation of equal terms of opposite sign yields

(3-11)
$$\Delta G_{T,P} = \mu\,\Delta n + \Psi\,\Delta q + f\,\Delta L.$$

In particular, when only work of mass transport is considered, the free-energy change for a reversible process at constant temperature and pressure is

(3-12)
$$\Delta G = \mu\,\Delta n,$$

in which Δn is, as before, the number of moles of the substance which enter a system and μ is the chemical potential in that region of space. When several processes are taking place at the same time, the total free energy is obtained by adding up the individual free-energy changes:

(3-13)
$$\Delta G_{T,P} = \mu_1\,\Delta n_1 + \mu_2\,\Delta n_2 + \mu_3\,\Delta n_3 + \cdots.$$

as should be for any extensive property.

3-11. Example

What is the change in free energy when Δn moles of a chemical species move from a region in which the species is at potential μ_1 to another region in which the substance is at potential μ_2, as depicted in Fig. 3-8?

The change in free energy for this process may be calculated in two steps. We compute first the change in free energy caused by the

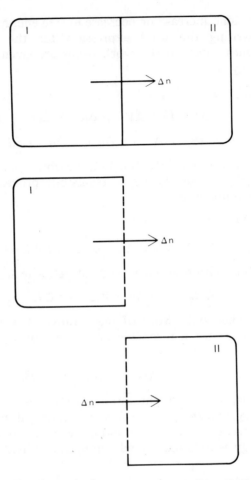

Fig. 3-8. The change in free energy when matter moves across a boundary can be calculated as the sum of the individual changes in each compartment; thus, $\Delta G_{total} = \Delta G_1 + \Delta G_2$.

entry of Δn moles into region 2, and then the change in free energy caused by the exit of the same Δn moles from region 1. These changes are:

$$\Delta G_1 = -\mu_1 \, \Delta n$$

and

$$\Delta G_2 = +\mu_2 \, \Delta n,$$

respectively. Notice that the signs have been chosen as negative and positive, respectively, depending on whether matter is leaving or entering the given region. The overall change in free energy becomes

(3-14)
$$\Delta G_{total} = \Delta G_1 + \Delta G_2$$
$$= \mu_1 \, \Delta n_1 + \mu_2 \, \Delta n_2$$
$$= - \mu_1 \, \Delta n + \mu_2 \, \Delta n.$$

Or, taking the number of moles outside a parenthesis,

(3-15)
$$\Delta G_{total} = (\mu_2 - \mu_1) \, \Delta n.$$

We can consider the following three cases:

- At equilibrium $\Delta G = 0$, so $\mu_2 = \mu_1$.
- In a spontaneous process, on the other hand, the free-energy change is negative—that is, free energy decreases. Since Δn is positive, the difference must be negative: This implies that $\mu_1 > \mu_2$.
- When the free-energy change is positive (larger than zero), no

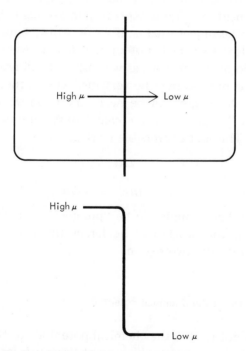

Fig. 3-9. During a spontaneous process, matter flows from a region of high chemical potential to a region of low chemical potential. This is the direction in which entropy increases and, at constant temperature and pressure, free energy decreases.

process takes place spontaneously because the final free energy would be larger than the initial free energy, thus violating the Second Law. In that case we see that $\mu_2 > \mu_1$.

We can conclude from the cursory analysis of these three extreme cases that the chemical potential μ acts like a height indicating the direction of the flow of matter in natural processes: Matter will flow from the high potential to the low potential (Fig. 3-9) in the same way that an object falls from a high to a low position in free fall on the earth's surface.

3-12. Can Mass Ever Move from a Low Potential to a High Potential?

We can ask the same question about an object in the gravitational field of the earth: Can an object move from a low to a high point? Clearly, this transition will not happen spontaneously, but it can take place if we lift the object; that is, if we do work. Similarly, mass will move from a low to a high potential region if work is expended. Since we can do as much work as we wish and still not transport the mass across (our process may be very inefficient), there is no way to assure how much energy will be necessary to move matter with a given procedure. We can, however, calculate the *minimum* amount of work required. This is the reversible work given by

$$(3\text{-}16) \qquad W_{rev} = -\Delta G_{T,P}$$

$$= (\mu_1 - \mu_2)\,\Delta n.$$

Then if the number of moles to be "pumped" to the high potential region is fixed, the work will be larger the larger the potential difference between the two regions.

3-13. Explicit Form of the Chemical Potential

So far we have not related the chemical potential—or the free energy, for that matter—to measurable quantities such as pressure and concentration. When the temperature, pressure, and electrical potential are fixed and the only variable is the amount of mass in various regions, it is clear that the more closely packed the mass is the harder it will be to add more of that species to the given region.

Packing, however, is not a function of the *total* amount of matter but, if matter is evenly distributed, a function of the amount of matter *per unit volume*. This quantity is called the *concentration* of the substance in a given region. Since usual concentrations contain too many molecules per unit volume, they are expressed in terms of moles per unit volume. A mole contains 6×10^{23} molecules of the substance. The actual relationship between concentration and chemical potential cannot be derived exactly; intuitively, however, we can see that the higher the concentration the higher the tendency of the molecules to leave the region. The question is *how much* will the potential increase when the concentration increases. Will the potential double and triple, for example, when the concentration doubles and triples? This particular type of change would be called linear and is shown in Fig. 3-10(a).

Actually, the typical concentration dependence in dilute solutions has the form shown in Fig. 3-10(b). It can be seen that with increasing concentration the actual increase in potential for a given concentration difference decreases. For example, the change in potential obtained when the concentration changes from 2 to 4

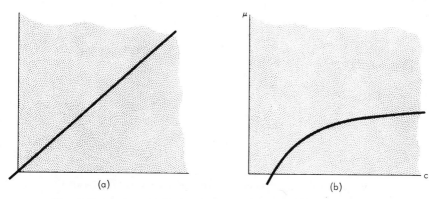

(a)   (b)

Fig. 3-10. If the chemical potential were linearly related to the concentration of a solute, its value would double every time the concentration was doubled, as shown in (a). Actually, the relationship is logarithmic and smaller changes are obtained, for the same variation in concentration, as the concentration increases (b).

molar is larger than that obtained when the concentration changes from 12 to 14 molar even though the concentration change is the same. This type of curve is called *logarithmic*.

For small changes in concentration, the change in chemical potential $\Delta\mu$ is experimentally found to be

$$(3\text{-}17) \qquad\qquad \Delta\mu = RT\frac{\Delta c}{c}$$

which clearly has the property that a small change Δc taking place at a high average concentration c will have a smaller effect on the chemical potential than the same change Δc occurring at a low average concentration c.

The usual way of writing the relationship between concentration and chemical potential is as

$$(3\text{-}18) \qquad\qquad \mu = RT \ln c$$

in which $\ln c$ is called the natural logarithm of the concentration. Equation 3-18 contains the same information as Eq. 3-17. Although we shall not regularly use Eq. 3-18, some of the relevant elementary properties of natural logarithms are reviewed in Fig. 3-11.

Potential differences, like differences in height, are relative, so the "zero" level must be set arbitrarily for a given temperature and

---

DIFFERENCE

If $\quad y_1 = \ln x_1 \quad$ and $\quad y_2 = \ln x_2$

then $\quad y_1 - y_2 = \ln\dfrac{x_1}{x_2}$

INVERSE FUNCTION

If $\quad y = \ln x$

$\quad x = e^y$

in which $\quad e = 2.7281\ldots$

LOGARITHM OF UNITY

$\ln 1 = 0$

LOGARITHMS OF NUMBERS LARGER AND SMALLER THAN 1

If $\quad x > 1, \quad$ then $\quad \ln x > 0$ (positive)

If $\quad x < 1, \quad$ then $\quad \ln x < 0$ (negative)

Fig. 3-11. Some elementary properties of natural logarithms.

pressure. The same occurs when we climb stairs: We do the same work by climbing three steps at the bottom as at the top of the ladder—even though people who suffer from dizziness may think otherwise. It is immaterial, in any case, where we start.

The logarithmic expression for the potential can then be written as*

$$(3\text{-}19) \qquad \mu = RT \ln c + \mu^0$$

in which μ^0, the standard chemical potential, specifies the "zero" level at a given T, P.

3-14. Complete Form of the Chemical Potential

The difference in chemical potential between two regions kept at concentrations c_1 and c_2 can now be given as

$$(3\text{-}20) \qquad \mu_1 - \mu_2 = (\mu_1^0 - \mu_2^0) + RT \ln \frac{c_1}{c_2}.$$

If the two regions are at the same temperature and pressure, the expression reduces to

$$(3\text{-}21) \qquad \mu_1 - \mu_2 = RT \ln \frac{c_1}{c_2}$$

because the standard chemical potentials μ_1^0 and μ_2^0 are only functions of temperature and pressure. In this case, if the concentration c_1 is larger than c_2, the ratio c_1/c_2 is larger than 1, the logarithm is positive, and μ_1 is larger than μ_2. When c_2 is larger than c_1, the logarithm is negative and μ_2 is larger than μ_1. If the two regions are not at the same pressure and temperature, the differences in these quantities also contribute to the chemical potential. It can be shown that the pressure term contributes a portion,

$$\overline{V} \, \Delta P$$

to $\Delta\mu$; in this expression, ΔP is the difference in pressure between the two sides—$\Delta P = P_2 - P_1$—while $\overline{V}$ is a quantity called the partial

---

*This equation actually stands for the difference

$$\mu - \mu^0 = RT \ln c - RT \ln c^0,$$

in which c^0 is set to 1 Molar.

molar volume, defined as the volume occupied by 1 mole of the given solute in 1 mole of solution when other variables are not changed:*

$$\text{Partial molar volume} = \frac{\text{volume of one mole of solute in solution}}{1 \text{ mole of solution}}.$$

The difference in chemical potential at constant temperature then becomes

$$\Delta\mu = \overline{V}\,(P_2 - P_1) + RT\,\ln\left(\frac{c_2}{c_1}\right)$$

and it is seen that when the concentrations are equal, the side at higher pressure will show the higher chemical potential and molecules will be "pushed" away from the region of higher pressure to the region of lower pressure. Similarly, the temperature contribution to the chemical potential difference is given by

$$-\overline{S}\,\Delta T,$$

in which ΔT is the temperature difference between the two compartments, while $\overline{S}$ is the partial molar entropy or entropy contributed to the total entropy by 1 mole of the solute in solution. The total chemical potential when pressure, temperature, and concentration effects are included is then given by

(3-22) $$\Delta\mu = \overline{V}\,\Delta P - \overline{S}\,\Delta T + RT\,\ln\left(\frac{c_2}{c_1}\right).$$

This formula has very useful biological applications which will be considered in the next section.

3-15. Ions and the Electrochemical Potential

Most molecules of biological importance have charges associated with them; these charges are not flying around by themselves but are attached to matter (molecules). Therefore, when charged matter moves, it is necessary to consider the electrical interactions with

---

*The volume occupied by the solute in solution is not the volume occupied by 1 mole of the solute before dissolving it. If n moles of H_2O are mixed with n moles of a solute, for example, the total volume of the solution is

$$V_{\text{total}} = \overline{V}_{H_2O}\,n_{H_2O} + \overline{V}_{\text{solute}}\,n_{\text{solute}}.$$

regions surrounding the charged molecule; the chemical potential, then, has to be modified to include electrical interactions. Charges attached to molecules have the same macroscopic properties as other charges:

- Charges are conserved; they can neither be created nor destroyed, and charge separations do not occur spontaneously on a macroscopic scale unless large forces are involved.

- Charges of equal sign repel each other, while different charges attract each other. Thus, it is necessary to do work in order to introduce charges in a region in which charges of the same sign are present.

As we have indicated before, the electrical work required to move a (positive) charge q into a region at electrical potential Ψ is

$$W = \Psi q.$$

Suppose a positive charge q is transported from a region at electrical potential Ψ_1 to a region at electrical potential Ψ_2. What is the work done? Referring to Fig. 3-12, we see that the work will consist of two terms: the work required to introduce the charge in region 2 minus the work done by region 1 on the charge; that is,

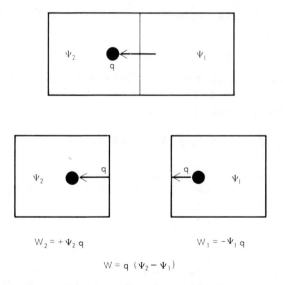

Fig. 3-12. The work done when a charge q is transported from a region held at potential ψ_1 to a region held at potential ψ_2 can be broken up into two terms.

$$W_{net} = -q\Psi_1 + q\Psi_2$$

$$\text{(3-23)} \qquad = q(\Psi_2 - \Psi_1).$$

Each charged molecule (ion) carries a charge denoted by z. Since charges come only in multiples of the electron charge, $e = 1.6 \times 10^{-19}$ coulombs, the charge per molecule is $e \times z$. The charge in 1 mole is this quantity multiplied by the total number of molecules in a mole, Avogadro's number, $N_0 = 6 \times 10^{23}$. The total charge in n moles is then

$$q = e \times N_0 \times z \times n.$$

This expression can be rewritten, after replacing $e \times N_0$ by F as

$$\text{(3-24)} \qquad e = F \times z \times n$$

in which z is the number of charges per ion, n is the number of moles, and F is the faraday—96,500 coulombs per charge per mole. z must be included with the proper sign; as an example, z for Ca^{++} is $+2$, for Cl^- is -1, for Na^+ is $+1$.

3-16. The Electrochemical Potential

If the electrical work per mole is introduced in the expression for $\Delta\mu$, we obtain:

$$\Delta\mu_{total} = \Delta\mu + \Delta\mu_{electrical} = \Delta\mu + q(\Psi_2 - \Psi_1).$$

The total potential difference is called the electrochemical potential and is written as $\Delta\tilde{\mu}$. The complete potential difference for the motion of charged molecules is, then,

$$\text{(3-25)} \qquad \Delta\tilde{\mu} = \overline{V}\,\Delta P - \overline{S}\,\Delta T + RT\,\ln\left(\frac{c_2}{c_1}\right) + zF\,\Delta\Psi.$$

This expression is, of course, also valid for uncharged matter since z becomes zero in that case and the formula reduces to Eq. 3-22.

Problems

3-1. A logarithm is always the logarithm of a number, *not* a number with units; what is, then, the meaning of writing $\mu = RT\,\ln\,c$, in which c has a unit as well as a number?

3-2. The enthalpy of a system, H, is defined as

$$H = E + PV.$$

(a) Show that enthalpy is a function of state; that is, that the same change in H is obtained for two different paths.

(b) What quantity does the enthalpy change give for an expansion at constant pressure?

3-3. We now prove the statement of the Second Law given by Clausius using the entropy principle. Consider a simple system divided into two parts, A and B. Let A be a bath at constant temperature T_A and B a bath at constant temperature T_B and suppose that a certain amount of heat flows from A to B spontaneously—that is, through an irreversible process. Can the fluctuation in S tell whether T_A is larger than T_B or T_B larger than T_A?

3-4. Starting with Eq. 3-25, derive the *explicit* change in free energy for 1 mole of a substance moving between two regions held at concentrations c_1 and c_2 for that species when no other species are present. Assume that the molecules are uncharged and that the two regions are at the same temperature and pressure. What are the equilibrium concentrations?

3-5. *The change in free energy for a simple chemical reaction.* Show that the change in free energy for the simple chemical reaction

$$A \longrightarrow B$$

when 1 mole of B is formed at constant T, P is given by

$$\Delta G = \Delta G^0 + RT \ln K_{eq}$$

in which ΔG^0, the standard free-energy change for 1 mole of the reaction is

$$\Delta G^0 = \mu_B^0 - \mu_A^0$$

and K_{eq} is the ratio of the equilibrium concentrations of B and A,

$$K_{eq} = \frac{c_B}{c_A}.$$

Hint: The reaction consists of two parts: the entry of B and the exit of B in the same region in space.

3-6. Refer to the equation given in the previous problem and answer the following questions:

(a) What is the free-energy change ΔG at equilibrium?

(b) What is the relationship between ΔG^0 and K_{eq}?

(c) If the standard free-energy change ΔG^0 is negative, will there be more or less A molecules than B molecules at equilibrium? If it is positive? (Recall that the logarithm of a number that is smaller than 1 is negative and the logarithm of a number larger than 1 is positive.)

3-7. Generalize the results above and obtain the free-energy change for a reaction like

$$aA + bB + cC \longrightarrow mM + nN + uU,$$

in which the small letters represent number of moles of the various chemical species.

3-8. Show that at constant temperature and pressure only electrochemical work is done.

3-9. Show that the free-energy change at constant temperature and pressure is given by

$$\Delta G_{T,P} = T \Delta_i S - W',$$

in which W' is the "useful" work. Does G increase or decrease during an irreversible process?

3-10. Show that the maximum amount of work that can be obtained from a process at constant temperature and pressure is equal to the free-energy decrease in the system.

3-11. Show that in a spontaneous (irreversible) process, the work output is less than the free-energy decrease of the system.

3-12. Many books state that the entropy change during an irreversible process is larger than the entropy change during a reversible process. How is this view reconciled with our statement that entropy is a function of state and its change is independent of the path?

4

MICROSCOPIC INTERPRETATION OF THERMODYNAMIC QUANTITIES

4-1. But What Is Entropy?

We now have a general idea of what the function S tells about natural processes and their directionality. The question still remains as to why entropy behaves the way it does and, for that matter, what entropy *is*. To look for an answer, we must leave the realm of the very large, the observable macroscopic changes in functions of state, and turn our attention to the microscopic motion of molecules, in the world of the very small.

Thermodynamics cannot deal with microscopic phenomena; it is meaningless, for example, to ask within the thermodynamic framework what entropy or internal energy is. Moreover, thermodynamics gives no information as to the detailed mechanisms that bring about changes in functions of state. Such considerations require microscopic models that make specific assumptions about molecular structure and intermolecular interactions.

Obviously, different models should be used to describe different systems. For example, there is no *a priori* reason to assume that water molecules will have the same kind of interaction whether they are in the form of steam, ice, or liquid; and, furthermore, one would have every reason to suspect that even small amounts of substances dissolved in pure water would alter its physical properties. Although all available microscopic models are gross oversimplifications of the real world, they can be useful provided they meet the following two minimal requirements:

- A microscopic model must be *self-consistent*—it must not show internal contradictions of any kind.
- It must be *consistent with the macroscopic description*—there must be an equation relating the microscopic and the macroscopic worlds. If this condition cannot be met, the model is useless.

4-2. Ideality and Molecular Cohesion

The simplest molecular models correspond to systems in which the molecules or atoms either interact with one another but do not move, or move but do not interact.

The first case corresponds to a *crystal*. A crystal can be viewed as a matrix in which atoms occupy fixed positions and are allowed only to vibrate (Fig. 4-1). The second model corresponds to an "ideal" gas, which obeys the equation of state

$$PV = nRT,$$

in which P = pressure
V = volume
T = temperature
R = the "universal" gas constant*
n = number of moles.

The only possible way for molecules to move throughout a container without interacting is for the molecules to have zero volume. In this rather unrealistic situation, they can go *through* each other without detecting a collision. An ideal gas cannot be bought in any store, but there are real gases that can approximate the behavior of an ideal gas; the lower the boiling point of a real gas, the closer its behavior will be to that of an ideal gas. Thus, steam is far from being ideal (it boils at 100° C); helium, on the other hand, is almost ideal (its boiling point is $-269°$ C).

$*R = 1.99 \dfrac{\text{cal}}{°\text{K-mole}} = 0.083 \dfrac{\text{liter-atm}}{°\text{K-mole}} = 8.35 \dfrac{\text{volt-coul}}{°\text{K-mole}}$

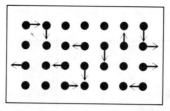

Fig. 4-1. In a crystal, each atom occupies a relatively fixed position and can only vibrate around it.

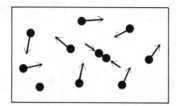

Fig. 4-2. In a gas, molecules can move relatively large distances before colliding.

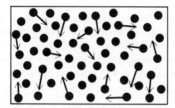

Fig. 4-3. In a liquid, molecules show strong neighboring interactions but can move in every direction.

Real gases show interactions that must be accounted for in terms of finite, rather than zero, volume for the molecules (Fig. 4-2). In a liquid, molecules show reasonably large interacting forces but they are still free to wander about the container, as shown in Fig. 4-3.

Since the ideal gas is the easiest system to treat, we shall restrict the following discussion to this model. Although very elementary, the model provides some insight and a semi-intuitive feeling for the physical basis of energy, entropy, temperature, and heat. Furthermore, most of the basic properties of ideal gases are shared by dilute solutions.

4-3. Ideal Gas Kinetics

A reasonable microscopic description of an ideal gas is given by a billiard-ball model with the following characteristics (refer to Fig. 4-4):

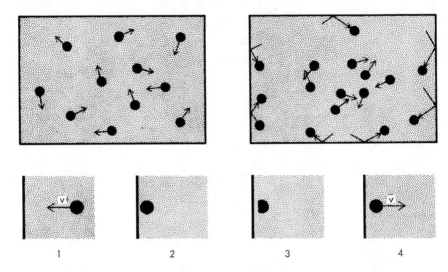

Fig. 4-4. In an ideal gas, molecules move in every direction; they occupy no space (so no collision occurs between molecules); and no energy is lost when a molecule collides with the walls of the container. The lower figures (1, 2, 3, 4) show the elastic interaction between a molecule and the wall at different times.

- Molecules occupy no space but have a finite mass, so their mass can be considered to be concentrated at a point.
- Molecules move at random in every possible direction.
- Molecules do not see each other but see the walls of the containers.
- Interactions with the walls are perfectly elastic—no energy is lost by a molecule that hits a wall.

As they stand, the ideal gas assumptions provide little information about the behavior of the system. The most important piece of the puzzle was contributed by Ludwig Boltzman, who assumed not only that molecules move in every possible direction but also that they move at the same speed. Later, Maxwell showed that the actual situation is one in which molecules in a gas can have *any* velocity, from zero to infinity, but they cluster around a most probable speed, as shown in Fig. 4-5.

In principle, one could write the equations of motion for each of the molecules in the container. While these equations can be given, an analytical solution—one that is given in terms of general variables rather than numbers—cannot be found when there are more than two molecules. The problem of finding a mathematical solution when only *three* bodies interact with one another has not yet been solved.

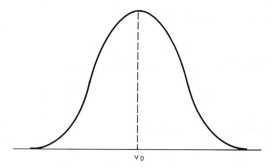

v_0

Fig.4-5. Molecular velocities in an ideal gas are distributed according to a bell-shaped, or Gaussian, curve. The peak corresponds to the number of molecules present which have the most probable speed in the container.

How many molecules are there in the volumes of mass we usually deal with in real life? Consider a volume of 22.4 liters (about 5 gallons) of any gas. In this volume, there are 6×10^{23} molecules; that is, 6 followed by 23 zeros. This number is very large and outside the realm of our experience; in real life we rarely come across numbers smaller than 0.01 (e.g., a cent) or larger than 1,000,000 (e.g., a million dollars). E. H. White gives the following analogy in his interesting book *Chemical Background for the Biological Sciences:*

> Suppose that we had 6×10^{23} peas; what volume would they occupy? Assuming that the peas were average in size (about 100 or 10^2 peas per cubic inch), how many peas would completely fill a household refrigerator? About 10^6 (one million). How many would fill an ordinary house from the cellar to attic? 10^9 (one billion). How many would be required to fill all the houses in a city the size of Chicago? 10^{15}. How many would be required to form a uniform layer 10 feet deep over the entire surface of the earth? 10^{22}. At this point, most of our peas still remain! To use up all the peas, we would have to blanket with 10 feet of peas about 60 planets the size of the earth!

The point is: If we cannot describe accurately the motion of three objects, how can we expect to be able to talk about the detailed motion of 6×10^{23} molecules we cannot even see?

A way out of this problem was given by J. Willard Gibbs. According to Gibbs, we only measure macroscopic phenomena that are averages of all the molecular processes taking place in the microscopic world; instead of worrying about detailed descriptions then, we concentrate on finding the *average* values of all quantities (energy, entropy, etc.) from statistical theory. This approach is called statistical mechanics and is one of the cornerstones of modern physics and cybernetics.

4-4. The Probabilistic Nature of Entropy

In Prob. 4-1 we will calculate the change in entropy when an ideal gas expands from a volume V_0 to a final volume V_1. This is given by

$$(4\text{-}1) \qquad \Delta S = S_1 - S_0 = nR \ \ln \frac{V_1}{V_0}$$

or, for a small change ΔV in volume, by

$$(4\text{-}2) \qquad \Delta S = nR \ \frac{\Delta V}{V_0}$$

when the expansion takes place at constant energy. Is it possible to reconcile this result with the microscopic model of a gas? Boltzman achieved this connection by assuming that entropy is related to the likelihood, or probability, of a given event's occurring in nature.

Consider an ideal gas in equilibrium. Molecules may be distributed in a number of ways in the box in which they are enclosed: They may be spread throughout the container, concentrated in the middle portion, etc. Of the many possible distributions, let us restrict our attention to the following two (Fig. 4-6): In the first case, molecules are evenly distributed; in the second, they are concentrated in the left side in a volume half of the total volume of the box. From everyday experience, we know that if the system starts in state 1, it will never go to state 2 spontaneously; and that if it starts in state 2, it will go to state 1. Furthermore, we have shown, from our previous calculation, that this occurs because according to the entropy principle the state with larger volume has a larger entropy and therefore the transition will occur in a direction which maximizes entropy.

Boltzman assumed that molecules in the ideal gas were moving at

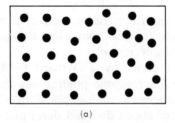
(a)

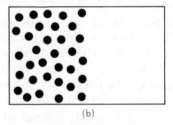
(b)

Fig. 4-6. Two possible arrangements of the molecules in a container. The more probable arrangement is that in which the molecules are distributed evenly (a) rather than unevenly (b).

random through the vessel in every possible direction, and although we cannot ask where a given molecule will be at a given time, we can determine the likelihood of finding a certain number of molecules in a given volume. Clearly, the larger the volume considered, the higher the probability of finding a certain number of molecules in the given volume. Suppose we ask what the chance, or probability, is of finding all the molecules inside the container. The answer is: absolute certainty; all the molecules will be there. This probability is arbitrarily denoted by $p = 1$. What is the probability of finding a molecule in a point of zero volume? Zero. The volumes in between will give probabilities between zero and one. This can be specified by the following equation:

$$(4\text{-}3) \qquad\qquad p = \frac{V}{V_{total}} ,$$

in which p is the probability of finding all the molecules in the container of volume V_{total} in some volume V (V is smaller than or equal to V_{total}).

If there is a transition from a volume V_0 to a volume V_1, the difference in probability between the two states is

$$(4\text{-}4) \qquad\qquad p_1 - p_0 = \frac{V_1 - V_0}{V_{total}} ,$$

which can be written, using the notation introduced for differences in Chapter 2, as

$$(4\text{-}5) \qquad\qquad \Delta p = \frac{\Delta V}{V_{total}} .$$

This equation is almost identical with the equation for the macroscopic entropy change (Eq. 4-2).

We can replace the volume change in Eq. 4-1 with the probability change given in Eq. 4-5 to obtain

$$(4\text{-}6) \qquad\qquad \Delta S = nR \ \Delta p,$$

which is also equal to

$$\Delta S = \frac{nR \ \Delta p}{p_{total}}$$

(since $p_{total} = 1$). We can then write

$$S = nR \ \ln \ p.$$

Boltzman gave the equation in the form

(4-7) $$S = k \ln p^*,$$

in which k is called Boltzman's constant. Since entropy is an extensive quantity, the entropy of N molecules must be N times the entropy of a single molecule,

(4-8) $$S = Nk \ln p,$$

or, if we consider a small change,

(4-9) $$\Delta S = Nk \ \frac{\Delta p}{p_{\text{total}}} = Nk \ \frac{\Delta p}{1} = Nk \ \Delta p.$$

Since this last result must agree with the macroscopic prediction, it follows by comparison with Eq. 4-6 that

(4-10) $$nR = Nk.$$

This equation provides a link between the macroscopic and microscopic worlds. Boltzman's constant has a value $k = 1.381 \times 10^{-16}$ ergs per $°C$.

4-5. Entropy, Probability, Order, and Disorder

Many books that do not introduce the statistical-probabilistic view of entropy refer to it as the amount of disorder in a system. To understand the relationship between order and probability, we consider an example borrowed from the theory of probability (although this particular example poses unsolvable philosophical problems for statisticians, it is perfectly appropriate when used in this context). Let a million monkeys sit in front of a million typewriters and ask for the probability that one of them will type Plato's *Republic*. Anyone who has ever met a monkey knows that the likelihood of this event's happening is very small; we would expect to get random and disordered series of letters and numbers, and the probability of the state with absolute order—in which letters are grouped together into meaningful words, words into sentences, and so on—is very small. Therefore, the most-ordered state, which is also the state of minimum entropy, is a state of low probability, while the least-ordered state (that of maximum entropy) is the state of highest probability.

---

*This formula is valid when looking for changes in entropy, but it does not tell us what the value of entropy is when $p = 1$ and the position of all molecules can be specified with certainty. Thus it should not be used to find values of entropy, only changes between states.

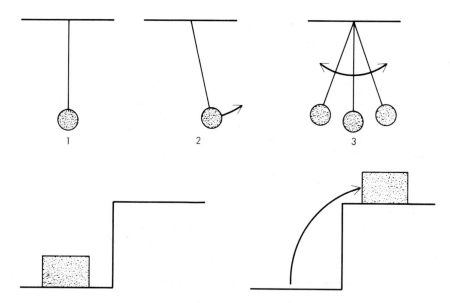

Fig. 4-7. A pendulum which starts to swing spontaneously and a block of steel which "jumps" provide examples of two events that have never been witnessed in the real world. Their occurrence would contradict the Second Law of Thermodynamics, although we would expect to see them happen if we waited for a long time—perhaps millions of years.

Whenever a statement of the Second Law of Thermodynamics claims that a certain process will not occur, it actually specifies that the likelihood of its happening is so small that in some cases, one would have to wait for long periods—sometimes millions of years—to witness the event (Fig. 4-7).

4-6. Relationship Between Microscopic Energy and Microscopic Entropy

A given system may exist in a number of possible energy states; a molecule, for example, may be vibrating, rotating, translating faster or slower, and as a result, its total energy—which is the sum of all these energy states—may acquire different values (Fig. 4-8). Classically, energy was believed to be changing continuously, so that between any two values one could always find another, intermediate value. Quantum mechanics has shown that this assumption is incorrect, as at the molecular and atomic levels energy changes by sudden jumps rather than continuously. In biology, as well as in physics, it is important to know the likelihood of a spontaneous

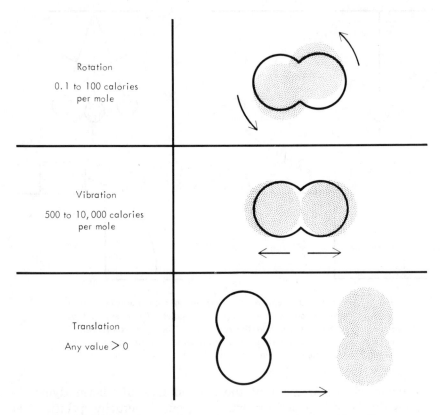

Fig. 4-8. Values of various forms of kinetic energy. (Adapted from N. Pimentel, "Chemical lasers," *Scientific American*, April 1965.)

jump between two energies, as this can answer the macroscopic question of what the probability is that a given biochemical process in the cell will suddenly change, or that a new molecule will be made spontaneously from a pre-existing molecule. For the purpose of fixing ideas, we consider a transition for a molecule that can exist in two different states, E_1 and E_0, and furthermore we shall assume that E_1 is larger than E_0:

$$(E_1).$$
$$(E_0) \nearrow$$

Suppose that the energy needed for the jump is provided by a constant temperature bath, at temperature T. The entropy of the bath will decrease by an amount,

$$\Delta S_{\text{bath}} = -\frac{Q}{T},$$

if the process takes place reversibly. The amount of heat, Q, transferred from bath to reaction vessel or container where the transition is taking place is given simply by

$$Q = E_1 - E_0,$$

since the increase in energy was achieved by transferring an equal amount of heat from the reservoir.

Equating the total change in entropy to the Boltzman formula we obtain,

$$\Delta S = k \ln p_1 - k \ln p_0$$

$$= k \ln \frac{p_1}{p_0}$$

$$= -\frac{(E_1 - E_0)}{T}$$

or, equivalently,

(4-11)
$$\frac{p_1}{p_0} = e^{-(E_1 - E_0)/kT} .$$

As the probability of finding a molecule in a given energy state is proportional to the total number of molecules in that state, it is clear that the ratio of the probabilities is equal to the ratio of the number of molecules in the two states, n_1/n_0. Equation 4-11 can then be written as

(4-12)
$$\frac{n_1}{n_0} = e^{-(E_1 - E_0)/kT},$$

which is plotted in Fig. 4-9. This equation, known as Maxwell-

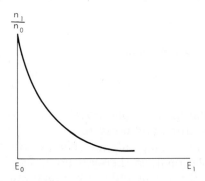

Fig. 4-9. The number of molecules with a certain energy decreases exponentially as we move away from the most probable energy level.

Boltzman's distribution, says that the number of molecules n_1 with a given energy E_1 decreases rapidly as we move away from the most probable energy E_0.

If all the energy is kinetic energy of translation, we can write

$$E_1 = \tfrac{1}{2} m v_1^2$$

$$E_0 = \tfrac{1}{2} m v_0^2,$$

in which v_0 is the most probable speed molecules in the container have, and v_1 is any other molecular speed. Maxwell-Boltzman's distribution then becomes

(4-13)
$$\frac{n_1}{n_0} = e^{-(1/2m)(v_1^2 - v_0^2)/kT}$$

It can be seen that although molecules can have any speed between zero and infinity the most probable speed is v_0. The *average* speed can be shown to be

(4-14)
$$v_{\text{average}} = \left(\frac{8\,kT}{\pi m}\right)^{1/2}$$

Typical speeds for gas molecules are given in Table 4-1 at $0°$ C.

TABLE 4-1

GAS	AVERAGE MOLECULAR SPEED (cm/sec)
Hydrogen	16.94
Water vapor	5.65
Carbon dioxide	3.62

4-7. Energy, Temperature, and Pressure of the Ideal Gas

We can now relate the microscopic model to macroscopic variables such as energy, temperature, and pressure.

Let us first assume that all the energy of the gas under consideration is translational, so there are no molecular vibrations or rotations. In practice, this is true only of monoatomic gases.

Refer to the drawing of Fig. 4-10. We are interested in finding

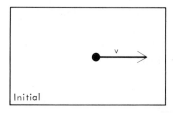
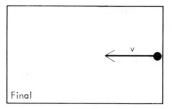

Fig. 4-10. The speed of a molecule is the same before and after collision with the walls of the container during an elastic collision.

the pressure applied by the molecules on the right wall. According to Newton's first law the force each molecule applies to the wall is equal and opposite to the force the wall applies to the molecule. If no energy is lost in the collision of the molecule against the wall, the velocity of the molecule *after* the collision has the same magnitude as it had *before* colliding with the wall but opposite direction. We can denote this situation by writing

$$v_{\text{final}} = -v_{\text{initial}}.$$

The *change* in velocity is, then,

$$\Delta v = v_{\text{final}} - v_{\text{initial}} = -2v_{\text{initial}}.$$

The force applied by the container wall to the molecule is given by Newton's second law of motion,

(4-15) $$F = m \times a,$$

in which a is the acceleration of the molecule during the time t the interaction lasts. The acceleration is given by

$$a = \frac{\Delta v}{t} = \frac{-2v}{t},$$

so the force applied to each molecule is, from Eq. 4-15,

$$F = \frac{-2v}{t} \times m.$$

The pressure applied by each molecule to the wall is, then,

(4-16) $$P = \frac{-F}{A} = \frac{2v}{At} \times m.$$

If $\vec{N}$ molecules with the same average velocity v hit the right wall in a time t, the total average pressure will be

(4-17) $$P = \frac{2vm\,\vec{N}}{At};$$

the question now is: How many molecules will hit the wall in that time?

We should first notice that in order for a molecule to reach the wall within t seconds, it must be within a distance

$$d = vt$$

or closer. All the molecules in the volume

$$V_1 = vtA$$

(shaded in Fig. 4-11) could in principle make it in time t *if* they were all moving from left to right. Actually, the N molecules in the container move at random, so they can go in six possible directions. The number of molecules that will go toward the right is then

(4-18)
$$\vec{N} = \frac{V_1}{V} \times \frac{N}{6} = \frac{vtA}{V} \times \frac{N}{6},$$

in which N is the total number of molecules in the container and V its (total) volume, so N/V is the density of the gas. We can introduce Eq. 4-18 in Eq. 4-17 to obtain

$$P = \frac{2v}{At} m \times \frac{vtA}{6} \times \frac{N}{V}$$

(4-19)
$$= \tfrac{1}{3} v^2 m \times \frac{N}{V}.$$

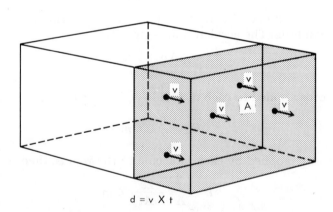

$$d = v \times t$$

Fig. 4-11. All the molecules that start within a distance $v \times t$ to the left of the wall will reach it in t seconds or less.

Since the kinetic energy of each molecule is

$$K.E. = \tfrac{1}{2} mv^2,$$

we can rewrite Eq. 4.19 as

(4-20) $$P = \tfrac{1}{3} \, [2 \, K.E.] \, \frac{N}{V}$$

in which K.E. is the average kinetic energy of each molecule in the container, or, rearranging terms,

(4-21) $$PV = \tfrac{2}{3} \, N \, [K.E.].$$

This formula must be equivalent to the macroscopic equation of state of an ideal gas,

$$PV = nRT$$

$$= NkT.$$

It then follows that

$$NkT = \tfrac{2}{3} N \, (K.E.) \quad \text{and}$$

(4-22) $$K.E. = \tfrac{3}{2} kT.$$

Thus, the temperature of an ideal monoatomic gas is a measure of the average kinetic energy of the molecules.

4-8. The Second Law and the Quality of Energy

While the First Law of Thermodynamics states that energy can be converted into other forms of energy, the Second Law says that not all forms of energy are equivalent. The ease of conversion from one type of energy into another depends on what the relative entropies of the two states are. As it turns out, each form of energy available in the universe has a characteristic entropy associated with it. This entropy depends again on how much "randomness" the given energy has.

Gravitational energy has the lowest entropy associated with it, and it is therefore a high-quality energy; heat, on the other hand, is low-quality energy. Table 4-2 gives the entropy associated with common forms of energy.

TABLE 4-2

FORM OF ENERGY	ENTROPY PER UNIT ENERGY
Gravitation	0
Energy of rotation	0
Energy of orbital motion	0
Nuclear reactions	10^{-6}
Internal heat of stars	10^{-3}
Sunlight	1
Chemical reactions	1-10
Terrestrial waste heat	10-100
Cosmic microwave radiation	10^4

Source: Reprinted from Freeman J. Dyson, "Energy in the Universe," *Scientific American*, September 1971.

Problem

4-1. Calculate the change in entropy, ΔS, when an ideal gas expands from a volume V_0 to a volume V_1 at constant temperature, T_0.

5

FREE ENERGY IN MOLECULAR BIOLOGY AND BIOENERGETICS

Since most biological processes occur at constant temperature and pressure, biological work must be accompanied by a corresponding decrease in free energy if the process occurs spontaneously; otherwise, the Second Law of Thermodynamics would be violated and we would be forced to postulate the existence of a vital force or supernatural agency. As we have seen in Chapter 3, free energy has the following important properties:

- G *decreases* in a spontaneous isolated process at constant temperature and pressure.
- The maximum amount of work a system can do on the environment equals the decrease in free energy.
- At equilibrium there is no change in G and free energy has its *minimum* value.

5-1. Relationship Between Standard Free-Energy Changes and Equilibrium Constants

The results of Prob. 3-7 can be used to calculate the change in free energy for a reaction in which two molecules, A and B, associate to form a compound AB according to the formula

$$A + B \longrightarrow AB.$$

The free-energy change (at constant temperature and pressure) was found to be given by

(5-1) $$\Delta G = \Delta G^0 + RT \ln K,$$

in which K is the ratio $K = [AB]/[A][B]$ and ΔG^0 is the standard free-energy change. It follows from Eq. 5-1 that the *total* free-energy change, ΔG, can be raised or lowered by changing the concentrations of A, B, and AB as shown in Fig. 5-1. In the particular case in which the reaction takes place *at equilibrium*, the total free-energy change ΔG is zero and we obtain, from Eq. 5-1,

(5-2) $$\Delta G^0 = -RT \ln K_{eq},*$$

in which $$K_{eq} = \frac{[AB]_{eq}}{[A]_{eq}[B]_{eq}}$$

and the concentrations of all species are now given at equilibrium. Taking into account the general properties of natural logarithms, it is possible to rewrite this expression as

(5-3) $$K_{eq} = e^{-\Delta G^0/RT}.$$

Equation 5-3 is drawn in Fig. 5-2; in general terms, the equilibrium constant decreases with increasing ΔG^0 at constant temperature. These changes are not simply proportional to the variation in ΔG_0 but much larger, as shown in Table 5-1.

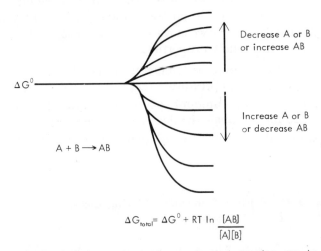

$$\Delta G_{total} = \Delta G^0 + RT \ln \frac{[AB]}{[A][B]}$$

Fig. 5-1. The free-energy change for a chemical reaction can be increased or decreased by changing the concentrations of reactants or products.

*Most books in biochemistry and molecular biology give this equation as a starting dogma without making it clear that it is valid only *at equilibrium*—which is not the most general situation encountered in living systems.

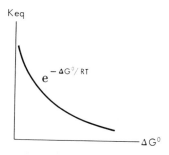

Fig. 5-2. The equilibrium constant *decreases* as the standard free-energy change *increases*.

The equilibrium free-energy change is usually given, throughout the biological literature, at a hydrogen concentration $[H^+] = 10^{-7}$ molar (pH = 7) and $25°$ C; this standard free-energy change is denoted as $\Delta G'$.

TABLE 5-1. Relationship Between Standard Free
Energy and Equilibrium Constants

K_{eq}	$\Delta G°$ (at $25°$ C)
0.001	4,089
0.01	2,726
0.1	1,363
1.0	0
10.0	−1,363
100.0	−2,726
1000.0	−4,089

Thus, in the simple reaction

$$A \longrightarrow B$$

if the free-energy change were −4,089, the equilibrium concentration of B would be 1,000 times larger than that of A, and if the free-energy change were −1,363, the concentration of B would be only 10 times larger than A.

In the case of the original reaction $A + B \longrightarrow AB$, if we start with 1 mole of A and 1 mole of B and the standard free-energy change is $\Delta G° = -4,089$ calories per mole, the reaction has essentially gone to completion at equilibrium.

5-2. Difference Between Direction and Rate

While thermodynamic theory predicts (Eq. 5-1) that chemical reactions which are accompanied by large decreases in standard free energy, ΔG^0, will have more product than reactant at equilibrium, it says nothing about *how long* it will take for the system to reach equilibrium. As a matter of fact, even though some reactions have large negative changes in standard free energy, it may take them *years* to reach equilibrium. This involves the microscopic mechanism (or path) of the reaction and is, therefore, outside the realm of macroscopic thermodynamic considerations.

Essentially, the problem arises because it is not enough that 2 atoms or molecules meet in a solution to form another molecule: They must collide with *enough* energy to stick to each other. This excess energy required over the standard free energy is called the *activation energy*. The situation is represented in Fig. 5-3: The system is in an initial state 1, which has a higher free energy than the final state 2. Thermodynamics assures, then, that the system will go spontaneously from state 1 to state 2 if we are willing to wait "long enough." But it is impossible to calculate what the activation-energy barrier is for a given reaction: This is *not a constant of nature*; we must have information as to the mechanism provided for the reaction to take place.

The natural question to ask is: How do we know that the transition will take place at all if the energy barrier is high? Recall that in any population of atoms or molecules at equilibrium, there is a range of possible velocities between zero and infinity; although most molecules have a velocity that is close to the average energy of the system, there are always a few that are way above this value and a few that are way below (Fig. 5-4). From time to time two reactant

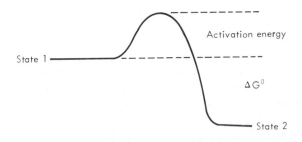

Fig. 5-3. For a process to occur at equilibrium, an activation energy must be provided in addition to the standard free-energy change. The activation energy varies from system to system.

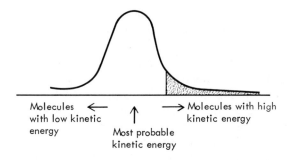

Molecules ⟵   ⟶ Molecules with high
with low kinetic   ↑   kinetic energy
energy   Most probable
              kinetic energy

Fig. 5-4. Transitions that require very high energy jumps will take place if we wait long enough, because there are always a few molecules with high kinetic energies (shaded area).

atoms or molecules with enough energy will meet and form a product (state 2). Thermodynamics says that once the transition to the more probable state takes place, the product will not tend to go back to the less probable state.

It follows from this discussion that unless the activation energy of a system can be lowered, most chemical reactions will take a long time to occur spontaneously, because the event in which two (or more) molecules with enough energy hit each other will rarely take place when activation energies are large.

One way of improving the chances of collision would be to hold one of the species at specific places in the solution. An even better way would be to provide specially shaped cavities in which both reactant molecules can fit within a close distance of each other, thus reducing the amount of kinetic energy required for the collision. This is the approach adopted by biological systems. The specially shaped cavities are found in special protein molecules, the *enzymes*. Most biochemical reactions involving small molecules are catalyzed—sped up—by enzymes. Each reaction has a specific enzyme associated with it, and it will not take place in a reasonable time unless the given enzyme is present.

Both enzymes and activation energies have a basic importance in maintaining life: If activation energies did not exist, all reactions would immediately go to their equilibrium states and living things would spontaneously degrade to reach the state of maximum entropy—completely random state. Enzymes, on the other hand, are needed to regulate the pace at which reactions take place as well as the sequence in which they occur.

5-3. Covalent vs. Weak Bonding: Molecular Biology
as the Chemistry of Molecular Recognition*

There is no apparent basic difference between the chemistry of small molecules and the chemistry of large molecules; the same rule of chemistry that applies to one kind applies to the other. There is, however, a logistic difference which distinguishes small atomic associations from larger ones.

Classical organic chemistry deals primarily with small organic *molecules.* A molecule is an association of two or more atoms held tightly together by *covalent bonds.* In the primitive "planetary model" of the atom the formation of covalent bonds was viewed as resulting from the sharing of electrons by the outermost orbits of two atoms, which in this way reached the stable configuration of noble, inert gases like helium, neon, argon, krypton, xenon, and radon. The modern quantum mechanical model has replaced the planetlike electrons with "electron clouds," in which electrons have a finite probability of being found at a given place.

Both in classical and quantum mechanical chemistry, the chemical properties of elements—the way in which atoms will react with other atoms to form covalent bonds—are dictated by the outermost layer of electrons.

Macromolecules found in biological systems are polymers made up of small molecular units that attach to each other by covalent bonds. (Fig. 5-5 shows some examples of biological interest.) But the way in which macromolecules "react" *is not* by forming more covalent bonds between them. The central theme of molecular biology is that once large biological polymers are formed they are able to fold and adopt many shapes, or "conformations" (Fig. 5-6). Among the many possible conformations, each polymer always adopts *only one* specific shape, which is also the configuration of minimum free energy at equilibrium. Once a given macromolecule is formed, its principal function is to *recognize* other small and large molecules. In the case of enzymes, for example, the protein must recognize two small molecules. To perform efficiently the enzyme, once it recognizes the substrate or substrates, must release the products as soon as they are formed. Enzymes act quickly (some recognize as many as 10^6 molecules per second), but, as we have pointed out in Sec. 5-2, it takes a long time for a molecule to be

---

*A more detailed discussion of molecular biology is given by J. D. Watson in *The Molecular Biology of the Gene.*

Monomer	Polymer	Biological role
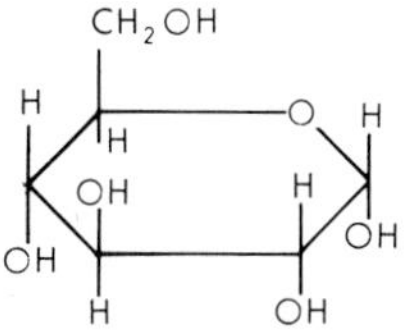 Glucose	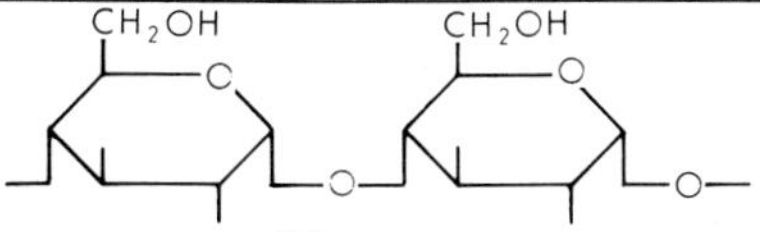 Glycogen	Long-term energy storage in animals
	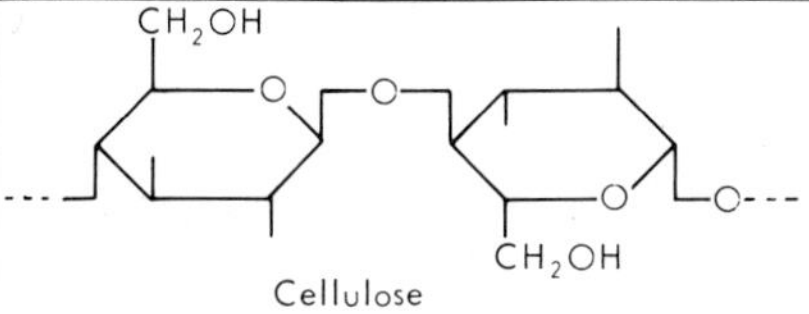 Cellulose	Long-term energy storage and structure in plants
Amino acids 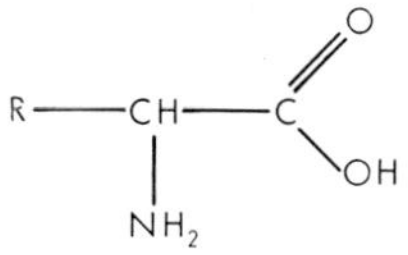	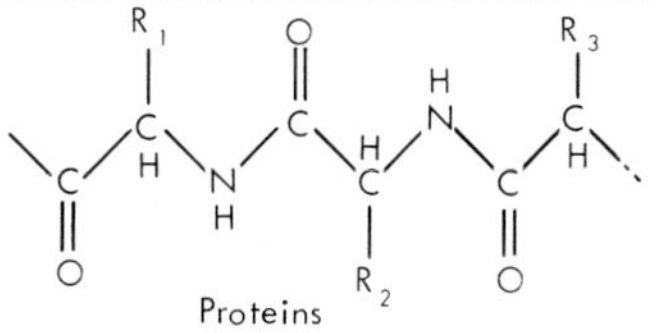 Proteins	Structure and function: Enzymes are proteins that catalyze specific biochemical reactions.
Deoxyribonucleotides 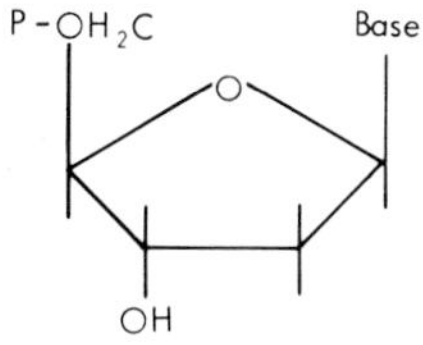	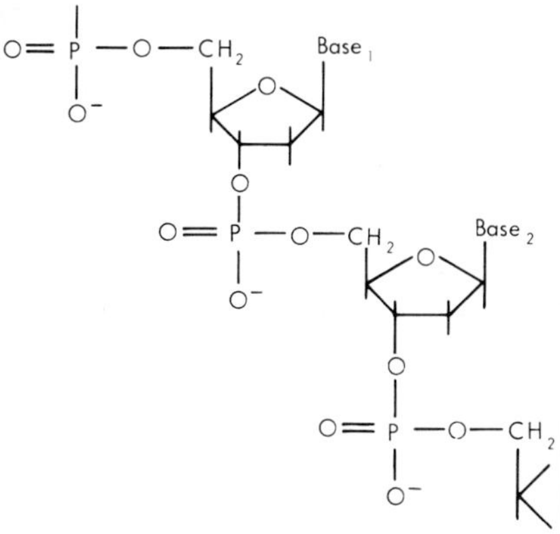	Long-term storage and transcription of genetic information in plants and animals
Ribonucleotides 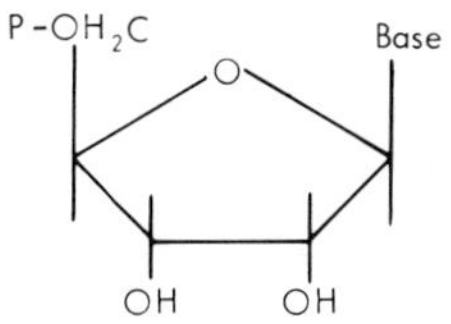	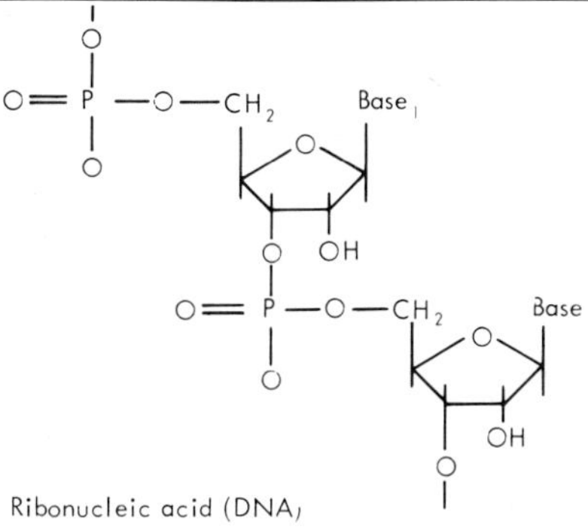 Ribonucleic acid (DNA)	Translation of genetic information into protein structure

Fig. 5-5.

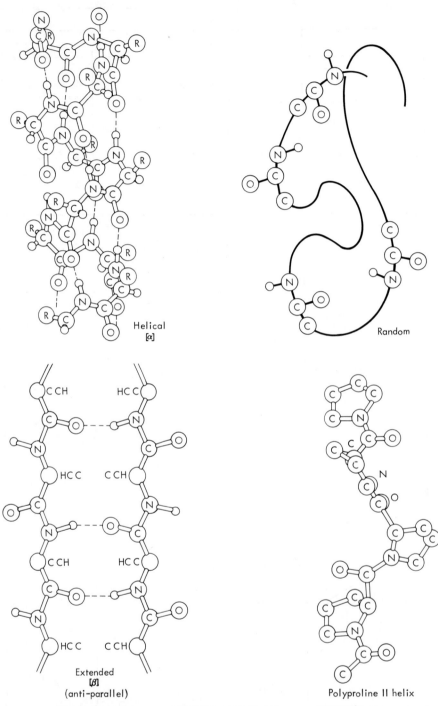

Helical
[α]

Random

Extended
[β]
(anti-parallel)

Polyproline II helix

Fig. 5-6. Some of the possible configurations a polypeptide can adopt. (E. R. Blout in *The Neurosciences*, ed. by G. C. Quarton. New York: Rockefeller University Press, 1967.)

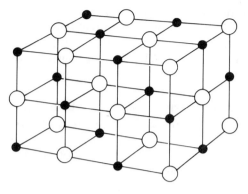

Fig. 5-7. A NaCl crystal consists of an even arrangement of Na^+ and Cl^- which alternate in space.

formed or destroyed in the absence of catalysts. It follows that the energies that hold the enzyme and the small molecule together must be much smaller than the covalent bonds energies that hold a molecule together. Indeed, while the standard free energy of covalent-bond formation from free atoms ranges between $\Delta G' = -50$ to -110 kilogram calories per mole, the typical standard free energy changes for substrate recognition—which can be calculated from equilibrium measurements—are of the order of 5 and 10 kilogram calories per mole. Other energies of "recognition" processes have also low values compared to covalent bonds.

5-4. What Is the Physical Basis of Weak Interactions?

There are several kinds of weak attractive and repulsive forces. All of them arise from electric interactions between atoms. The most basic one occurs between permanently charged atoms (ions). A crystal of sodium chloride (common salt), for example, has the general structure shown in Fig. 5-7: Na^+ and Cl^- ions alternate in a very regular (cubic) lattice and remain in position solely by the electrostatic attractions and repulsions between atoms. The attractive force acting between two charges of different sign (positive and negative) is given by Coulomb's law:

$$F = \frac{-q_1 q_2}{\epsilon r^2},$$

in which q_1 and q_2 are the charges, r their separation, and ϵ the dielectric constant, a property characteristic of the electrical properties of the medium where the charges are. The dielectric constant ϵ is 1 for vacuum and 81 for H_2O; as a result, while the ionic forces that hold a Na Cl crystal together are comparable to those of a covalent bond, the attraction between these ions in a water solution is rather weak.

A second type of electrical attraction which appears between any two atoms consists of the London dispersion forces, which are caused not by permanent charges but by induced charge fluctuations between two atoms. Unlike the ionic interactions which vary as the inverse square of the separation distance, l/r^2, London dispersion forces decrease very quickly as the inverse sixth power of the interatomic distance, l/r^6. This generalized attractive force is only effective at short intermolecular separations. Moreover, since the "bonds" formed are weak ($\Delta G' = 1$ to 2 Kcal/mole), only when several bonds are formed at once will two structures be bound. At very close interatomic distances, when the electron "clouds" of 2 atoms start to interpenetrate, there is also a repulsive force, the *Van der Waals'* force. London-Van der Waals' forces balance out at equilibrium, and the resultant separation distance, which is typical of each atom, is called the *Van der Waals' radius* (Table 5-2). The distances given in the table are larger than typical covalent bonds. The covalent-bond distance between H and O in H_2O, for example, is only 0.95 angstrom.

TABLE 5-2

ATOM	VAN DER WAALS' RADIUS (Å)
H	1.2
N	1.5
O	1.4
P	1.9
S	1.85
CH_3 group	2.0

Source: J. D. Watson, *The Molecular Biology of the Gene* (New York: W. A. Benjamin, 1965).

A third kind of weak bond which is very important in biological molecules is the *hydrogen bond*. This form of bonding, which is stronger than the Van der Waals' - London interaction but weaker than that between atoms covalently bound, occurs between a negatively charged atom and a hydrogen atom which is (covalently) bound to a third atom and which carries some positive charge. The hydrogen bond can be represented as $A-\ldots H^{+}-B^{-}$; the strength of the bond increases with increasing acidity of the group HB, but in the extreme case in which the charge increases excessively the H atom (proton) will be transferred to A^{-}, so the resultant bond is $A^{-}H^{+}\ldots B^{-}$.

5-5. Is the Hydogen Atom Charged?

This seems to oppose the statement that a covalent bond is formed by sharing electrons and that the overall charge in a molecule is zero. Actually, the total charge in a molecule is zero, but when the molecule is not symmetric there will in general be an uneven distribution of electrons, as some atoms will have more affinity for electrons than other atoms. Furthermore, this charge asymmetry has a distinctive directionality or polarity along the direction of the covalent bond $H - B$, so in order for the hydrogen bond to be formed between A and H, A must be aligned with the direction of the covalent bond $H-B$. Typical hydrogen bond lengths are given in Table 5-3.

TABLE 5-3

BOND	BOND LENGTH (Å)
OH . . . . O	2.70
OH . . . . O^{-}	2.63
OH . . . . N	2.88
NH . . . . O	3.04
NH . . . . O	2.93
NH . . . . O	3.10

Source: J. D. Watson, *The Molecular Biology of the Gene* (New York: W. A. Benjamin, 1965).

5-6. Activation Energies of Weak Bonds

Unlike the standard free energies for the formation of covalent bonds, the standard free energies for the formation of weak bonds are low and of the same order of magnitude as the typical average kinetic energies of heat motion (or only slightly larger). This insures, according to the previously discussed kinetic view of activation energies, that there will always be many molecules with enough energy to form (and break) weak bonds. Moreover, since the frequency of collisions must be proportional to the number of molecules with velocities above a certain minimum value, it is clear that the presence of many molecules with high energy assures that the *rate* at which weak bonds form and break is high.

5-7. If Weak Bonds Form and Break so Fast, Are There a Substantial Number Present at Equilibrium?

We stress again that there is a great difference between rate and direction. The fact that molecules in solution can impart enough energy per unit time to bonds to break them at a fast rate implies that they will have enough energy to "push" molecules in solution together to form new bonds also at a high rate. The number of bonds relative to the number of broken bonds—those that have the potentiality to form but have not—is dictated only *at equilibrium* by the magnitude and sign of the standard free-energy change. We saw in Sec. 5-2 that even a low negative free-energy change of the order of $\Delta G' = -2$ kilogram calories is enough to carry a reaction to completion. In the present context this implies that, at equilibrium, there will always be a large number of weak bonds formed.

5-8. The Importance of Three-dimensional Fitness

Since Van der Waals', hydrogen, and ionic bonds are so weak, molecular associations involving these bonds and having a certain degree of stability and specificity must involve many bonds per unit area of interaction between molecules. For this reason, macromolecules show cavities and protrusions which fit into other molecules in a way that has been compared to a lock-and-key mechanism. This close three-dimensional fitness assures that many secondary weak bonds can be formed. As representative examples, we can mention the double helical structure of DNA which is stabilized by hydrogen bonds, the active site of an enzyme which recognizes its substrate and the complementary three-dimensional regions of macromolecular aggregates which are made by association of several protein chains (Fig. 5-8).

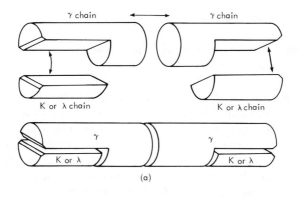

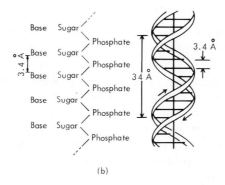

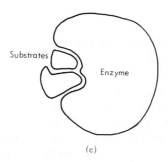

Fig. 5-8. Structural drawing showing the "three-dimensional fitness": (a) immunoglobulins (G. M. Edelman in *The Neurosciences*, ed. by G. C. Quarton. New York: Rockefeller University Press, 1967); (b) base pairing in DNA; (c) enzyme and substrates fit.

5-9. Three-dimensional Shapes of Macromolecules Are Configurations of Minimum Free Energy

Why do macromolecules—proteins, for example—have specific and individual three-dimensional shapes? Whatever the shape, the equilibrium free energy must have a minimum value at constant T, P. The forces that bring the molecule to its equilibrium configuration are hard to give in a theoretical model with predictive value because of the many possible configurations that can in principle be reached. However, the basic forces that shape macromolecules are easy to understand in the light of our previous discussion. Consider a protein; the apparatus for transcription of genetic information makes only a *linear* (flat) array of covalently bound amino acids of the form

$$AA_1 - AA_2 - AA_3 - \ldots - AA_n.$$

This relatively long chain can fold in many possible ways. If it could fold completely at random by rotating about any of its bonds, a protein consisting of 100 amino acids would have a large number of possible configurations. Actually, the protein chain is restricted in the number of ways it can fold. One of the main reasons is that the peptide bond is a rather rigid planar structure (Fig. 5-9) which cannot rotate. Moreover, rotation around other covalent bonds is hindered by the "bulkiness" of the amino acid side groups.

Once the linear protein folds, it forms some new covalent bonds between sulfhydryl groups (HS) and numerous hydrogen bonds which stabilize the structure. There are also attractive and repulsive electrostatic interactions between side groups which contribute to determining the final shape of the finished molecule.

5-10. How Biological Free Energy Is Obtained and Put to Work: Bioenergetics

The basic energy required for biological processes is provided by foodstuffs—which are storing, directly or indirectly, the energy that arrived from the sun. Weight-watchers are familiar with the energetic balance. When their booklets mention "so many calories per pound," for example, they refer to the free-energy decrease that is obtained by burning these foods in oxygen. Most foods, however, contain many different types of molecules such as proteins and carbohydrates; we are presently interested only in free-energy changes of very simple chemical reactions.

What are the typical free energies supplied by various foodstuffs?

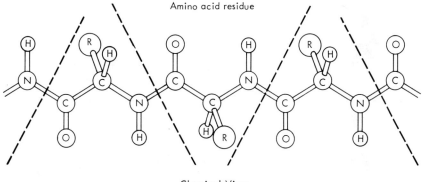

Amino acid residue

Chemical View

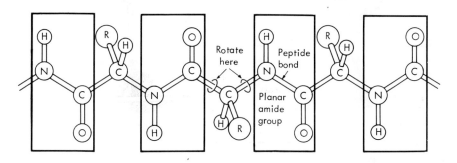

Structural View

Fig. 5-9. As the polypeptide backbone has a considerable amount of double bonding, rotation is restricted to the regions between the planar amide groups. (From Elkan R. Blout in *The Neurosciences*, ed. by G. C. Quarton et al. New York: Rockefeller University Press, 1967.)

The oxidation of glucose, a typical energy-providing molecule, proceeds according to the equation

$$\text{Glucose} + \text{oxygen} \longrightarrow CO_2 + H_2O$$

and has a standard free-energy change, in the direction written, of

$$\Delta G' = -686,000 \text{ cal/mole (180 g of glucose)}$$

at physiological conditions. Palmitic acid, a lipid, burns in oxygen with a free-energy decrease of

$$\Delta G' = -2,338,000 \text{ cal/mole.}$$

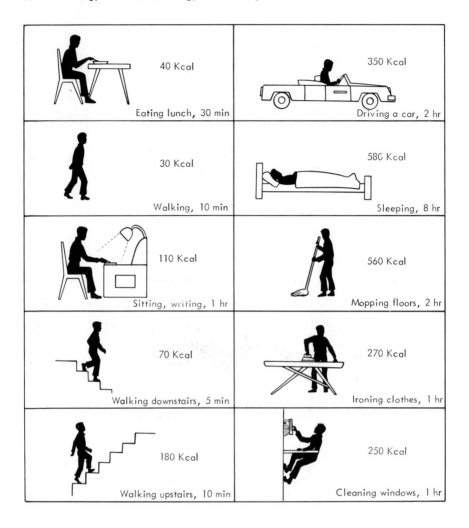

Fig. 5-10. Approximate metabolic-energy expenditure during various activities for a man weighing 70 kg.

We can compare these numbers with typical energies spent by the human body during various types of daily activities. This is shown in Fig. 5-10, from which we see that a 154-pound man walking upstairs *for an hour* would spend about 1,000,000 calories. Clearly, 686,000 calories is a large amount of energy. The amount of work that is actually done by the human body with this energy is much less than this, as not all the available free-energy decrease is transformed into work by irreversible processes. The actual efficiency is less than 40%. Moreover, foodstuffs are not burned directly in oxygen to release the energy directly in the form of heat; rather, their energy is released in steps while going through relatively complex sequential chemical transformations that release small "packages" of energy at a time.

5-11. The Recovery of Chemical Energy

We have already shown (Prob. 3-8) that at constant temperature and pressure the only useful work that can be done is electrochemical work. In particular, no heat can be used to do work under these conditions. For this reason, biological "machines" use chemical energy to do work. The energy of foodstuffs is transferred to molecules that have the ability to store it for either short or long periods and can release it at later times when work must be done. Notably among these specialized molecules stands adenosine triphosphate (ATP), which has the formula shown in Fig. 5-11. ATP hydrolyses—dissolves in water—to yield the simpler compound adenosine diphosphate (ADP, depicted in Fig. 5-12) and inorganic phosphate,

$$ATP + H_2O \longrightarrow ADP + P.$$

This reaction proceeds to the right with a relatively large standard free-energy change,

$$\Delta G' = -7 \text{ Kcal/mole (at } T = 25° \text{ C}, ph = 7)$$

$$= -7000 \text{ cal/mole.}$$

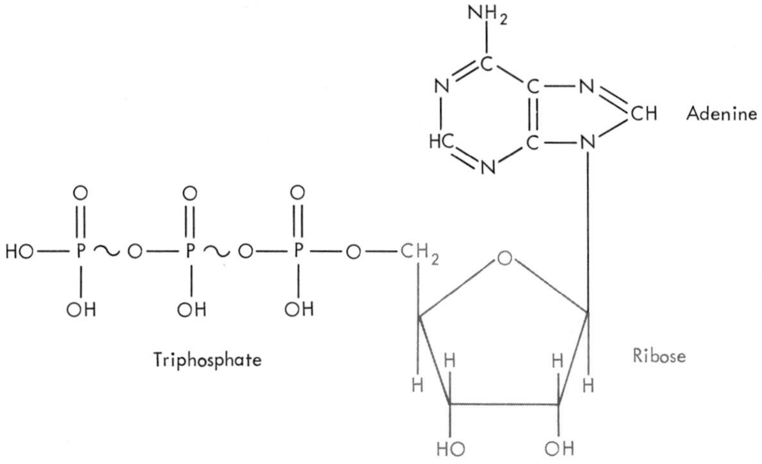

Fig. 5-11. The chemical structure of ATP.

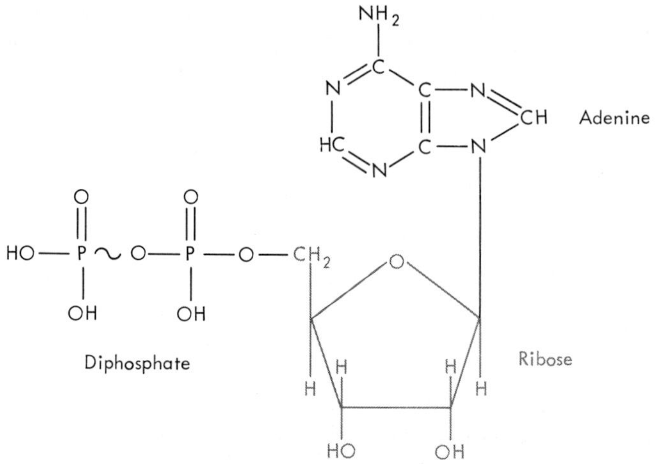

Fig. 5-12. The chemical structure of ADP.

Most biological work depends on the donation of the free energy of hydrolysis of ATP to energy requiring reactions. Since the appearance of ADP means that energy has been spent and the presence of ATP means that energy is available, we could think of ATP and ADP as the "charged" and "discharged" forms of the molecule. We should point out, however, that ADP can be further hydrolysed to yield adenine monophosphate (AMP), which can also be considered a high-energy molecule, although this second hydrolysis gives a smaller standard free-energy decrease. These molecules charge and discharge their energy so that they can be used over and over. How is ATP charged and discharged? In order to understand how the process works we consider the idea of chemical coupling.

5-12. Coupling Chemical Reactions

Suppose that we are given the reactions

$$A \longrightarrow B + C$$

and
$$C \longrightarrow D.$$

If we add them up, we obtain a new chemical equation

$$A + C \longrightarrow B + D + C$$

and since the species C appears both as a reactant and product it does not take part in the net reaction,

$$A \longrightarrow B + D.$$

Driving reaction

A ⟶ B

Driven reaction

C ⟶ D

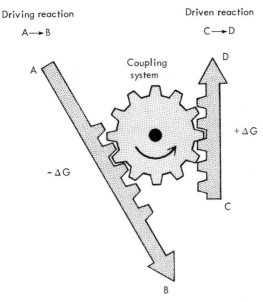

Fig. 5-13. If a chemical reaction has a large negative free-energy change, it can sometimes couple with an unfavorable reaction to drive it against the direction in which it would tend to go.

Suppose that the first reaction has a standard free-energy change of −8 kilogram calories per mole, and the second a free-energy change of +4 kilogram calories per mole at the given conditions of temperature and pressure. The first reaction can occur spontaneously but the second cannot. What about the net reaction? By adding their individual free energies when both reactions take place, we obtain +4 −8 = −4. Since the overall reaction has a negative free energy, it will occur spontaneously. Thus, if we put species C in a beaker, it will not go to species D spontaneously; however, if we add A, species D will be formed. This is a central point of biological energy transformations: *Chemical work is done by attaching reactions with a large negative free-energy change to reactions with unfavorable free-energy changes* (Fig. 5-13). We stress again that the negative free-energy change merely specifies the *direction* of the process and not the time it takes for the reaction to go to completion. In the absence of enzymes this may take forever.

The main requirements for driving a chemical reaction in a direction in which it would not spontaneously occur are then:

- There must be another reaction that proceeds with a larger decrease in free energy.
- The two reactions must share a common intermediate.

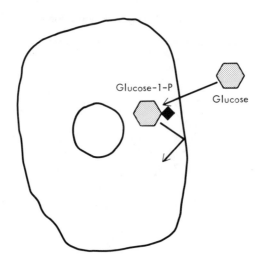

Fig. 5-14. Many molecules can go in and out of the cell freely if they are not in their "phosphate" form. When a phosphate group is added, they cannot pass through the cellular membrane.

The hydrolysis of ATP meets these two requirements: It proceeds with a decrease in free energy which is large enough to drive other reactions, and it can couple to other reaction by transferring the phosphate group. There are many small molecules in the cell which can be "charged" by adding a phosphate group (a common example is glucose) which can be activated to "high energy" from glucose 6-phosphate (Fig. 5-14). Phosphate compounds of small molecules also have the peculiar advantage that they cannot leave the cell, even though their "discharged" counterparts can pass freely through the cell membrane, as shown on the same figure; in this way, they cannot be lost before they are used for the required reactions.

Compared to the free energy of hydrolysis of other high energy compounds, the free energy of ATP hydrolysis has an intermediate rather than a very high value. Table 5-4 gives the standard free energy of hydrolysis of common "high energy compounds." It can be seen that ATP occupies an intermediate position between the very high energy compounds and the very low energy compounds and can, therefore, transfer energy from compounds whose free energy of hydrolysis is much larger to those with low free energy of hydrolysis. These low energy compounds can, then, be driven against the direction in which they would decompose spontaneously. For example, the reaction

$$\text{Glucose 1-phosphate} \longrightarrow \text{glucose} + \text{P}$$

which takes place in water proceeds, according to Table 5-5, with a standard free-energy change of

TABLE 5-4 Standard Free Energies of Hydrolysis
of Phosphate Compounds at pH = 7
and 25° C

	$\Delta G'$ CAL/MOLE
Phosphoenolpyruvate	−12,800
1,3-diphosphoglycerate	−11,800
Phosphocreatine	−10,500
Acetyl phosphate	−10,100
ATP	− 7,000
Glucose 1-phosphate	− 5,000
Fructose 6-phosphate	− 3,800
Glucose 6-phosphate	− 3,300
3-phosphoglycerate	− 3,100
Glycerol 1-phosphate	− 2,300

Source: A. Lehninger, *Bioenergetics* (New York:
W. A. Benjamin, 1965).

$$\Delta G' = -5000 \text{ Kcal/mole.}$$

It will then occur spontaneously as written. If a large enough source
of free energy is provided—such as the hydrolysis of ATP—the
reaction can run in reverse as

$$\text{Glucose} + P \longrightarrow \text{glucose 1-phosphate,}$$

even though it is unfavorable in the direction given ($\Delta G' = +5000$
cal/mole). The total free-energy change will, of course, be negative:

$$\Delta G'_{\text{total}} = -7 \text{ Kcal/mole} + 5 \text{ Kcal/mole} + -2 \text{ Kcal/mole.}$$

An important example of energy activation through phosphate
group transfer is the activation of amino acids. The reaction
schematically represented by

$$AA_1 + AA_2 \longrightarrow AA_1 - AA_2,$$

in which two amino acids join to form a dipeptide with a free-energy
change of 1 to 4 kilogram calories per mole. This means that two
amino acids in solution will not join spontaneously to form the
dipeptide: It is necessary to provide a source of negative free energy.
This is achieved by "activating" the amino acids to a high-energy
state, coupling the reaction

$$\text{ATP} \longrightarrow \text{ADP} + P$$

to $$\text{AA} + \text{ATP} \longrightarrow \text{AMP} - \text{AA} + P.$$

5-13. Other Forms of Biological Work
Also Require ATP Hydrolysis

While phosphate group transfers have an important role in coupling and driving chemical reactions such as the assembly of macro-molecules, the free energy of hydrolysis of ATP is utilized in many other forms of biological function requiring negative free-energy changes. Some of these processes involve the coupling of ATP hydrolysis to mechanical energy of contraction or motion (e.g., muscle, sperm tails), while others involve the coupling of ATP's chemical energy to transport solutes against their electrochemical potential gradients (e.g., active transport of Na^+ in nerve work, glandular secretion, absorption of metabolites by the intestine). Figure 5-15 illustrates a relevant example. Although the detailed discussion of postulated mechanisms of biochemical coupling belongs in biochemistry textbooks, it is relevant to point out that two general features of animal tissues performing biological work are the presence of a large number of mitochondria, the cellular organelle that manufactures ATP, and the presence of ATPase, the enzyme that catalyzes the hydrolysis of ATP. From the thermodynamic point of view, it is sometimes important to calculate the free-energy decrease required to perform a given type of work. This calculation cannot be undertaken exactly before the specific mechanism is known, but the *minimum* free-energy decrease must equal the reversible work done at equilibrium.

5-14. Conservation of Biological Energy:
The Generation of ATP

ATP is generated in a number of ways, depending on the organism. In some cases energy release during fuel burning (food breakdown) can be coupled to the synthesis of ATP from ADP and inorganic phosphate. In photosynthetic plants, on the other hand, radiant energy provides the required free-energy decrease.

We can consider the general features of the process of energy recovery in aerobic organisms as an example. The general formula for respiration,

$$\text{Food} + O_2 \longrightarrow CO_2 + H_2O + \text{energy} + \text{by-products},$$

has a deceiving simplicity. The overall process consists of many enzymatic reactions which can be summarized as shown in Fig. 5-16.

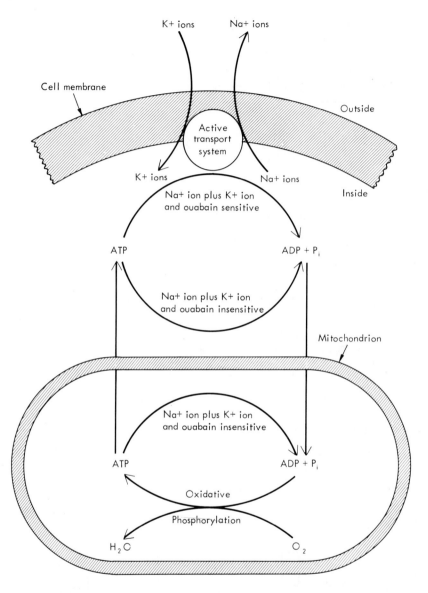

Fig. 5-15. Active transport is one of the biological processes which involve ATP hydrolysis to use chemical energy. (R. Whittam in *The Neurosciences*, ed. by G. C. Quarton. New York: Rockefeller University Press, 1967.)

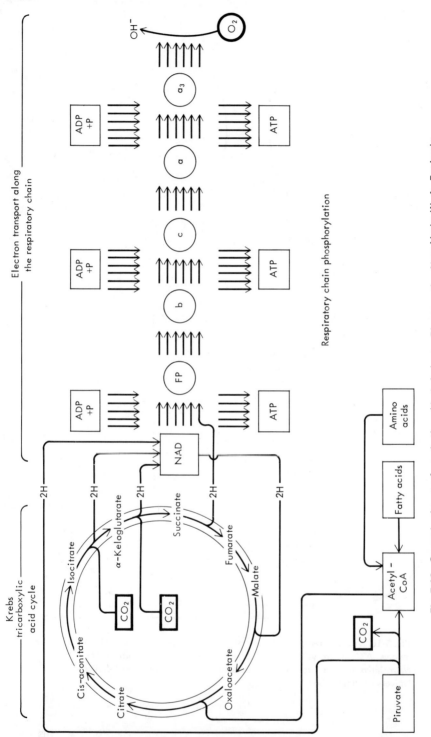

Fig. 5-16. General scheme of respiration. (A. Lehninger, *Bioenergetics.* New York: W. A. Benjamin, 1965.)

Foodstuffs (carbohydrates, fatty acids, and amino acids) first enter a series of reactions, collectively called the "Krebs tricarboxylic acid cycle," in which carbon skeletons are broken down into CO_2 molecules but no ATP is generated. The next step in oxidative phosphorylation is the transfer of electrons among a series of electron-accepting enzymes (respiratory chain enzymes). ATP is generated in these steps, and it is the final step in the electron transfer chain which gives an electron to molecular oxygen to form O^- and water. Typical ΔG's are given in Fig. 5-17.

Anaerobic microorganisms can generate ATP through the process of fermentation, and still other organisms—facultative—can undergo either fermentation or respiration, but in these cases respiration is a more efficient process.

Problems

5-1. Consider a cell that pumps in an uncharged solute whose concentrations inside and outside the cell obey the ratio

$$\frac{[\text{Solute}]\,\text{in}}{[\text{Solute}]\,\text{out}} = e^2.$$

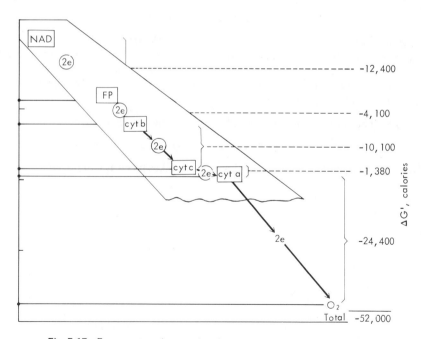

Fig. 5-17. Free-energy changes in the electron transport chain. (A. Lehninger, *Bioenergetics.* New York: W. A. Benjamin, 1965.)

(a) What is the minimum work needed to pump the solute? ($T = 25°$ C)?

(b) How much work must be done to pump a *charged* solute with $z = +1$ and the concentration ratio given above if the inside of the cell is 58 mV positive with respect to the outside?

(c) How many moles of ATP would be required if this process were perfectly efficient?

5-2. The reaction

$$\text{Glucose + fructose} \rightarrow \text{sucrose} + H_2O$$

has a standard free-energy change of $\Delta G^0 = +5500$ cal/mole. Will this reaction occur spontaneously under any conditions?

5-3. The reaction of Prob. 5-2 can be coupled to the hydrolysis of ATP to give the overall reaction

$$\text{ATP + glucose + fructose} \rightarrow \text{sucrose + ADP + phosphate}$$

with a standard free-energy change (at $25°$ C, pH 7) of $\Delta G' = -1,500$ cal/mole. What is the equilibrium constant for this reaction? Will it occur spontaneously under all conditions?

5-4. The assembly of the long-term energy-storing molecule glycogen from glucose monomers can be described by the equation

$$\text{Glucose + glycogen}_n \rightarrow \text{glycogen}_{n+1},$$

in which the subscripts refer to the number of glucose molecules in the given glycogen polymer. The standard free-energy change for this reaction is, at standard conditions, $\Delta G^0 = +5,000$ cal/mole. How many moles of ATP would be required to add 1 mole of glucose to 1 mole of glycogen? Assume the process is perfectly efficient and reversible.

5-5. The actual overall equation for the polymerization of glycogen is

$$\text{Glucose + 2 ATP + glycogen}_n \rightarrow \text{2 ADP + 2 P + glycogen}_{n+1}.$$

How many cal/mole are lost when the reaction takes place at equilibrium?

6

THE ELECTROCHEMICAL POTENTIAL AS A MEASURE OF BIOLOGICAL EQUILIBRIUM: OSMOTIC PRESSURE, NERNST AND DONNAN POTENTIALS

6-1. When Is the Electrochemical Potential a Measure of Equilibrium?

The general equilibrium condition at constant temperature and pressure is, as we have seen, that the change in free energy for the given process is zero; that is,

$$(6\text{-}1) \qquad \Delta G_{T,P} = 0 \text{ (at equilibrium).}$$

Moreover, the free-energy change under these conditions can also be written as

$$\Delta G_{T,P} = \mu_1 \Delta n_1 + \mu_2\, \Delta n_2 + \mu_3\, \Delta n_3 + \ldots$$

It follows that for the special case in which Δn moles of a substance go from a compartment in which the chemical potential is μ_2 to a compartment in which the chemical potential is μ_1, the change in free energy will be (Eq. 3-15)

$$(6\text{-}2) \qquad \Delta G = (\mu_2 - \mu_1)\Delta n.$$

This implies that at equilibrium $\mu_1 = \mu_2$. If the substance is charged, the equilibrium condition becomes

$$\tilde{\mu}_1 = \tilde{\mu}_2, \qquad \text{or}$$

(6-3) $$\tilde{\mu}_1 - \tilde{\mu}_2 = \Delta\tilde{\mu} = 0.$$

These equilibrium considerations are very important in several physiological problems in which we deal with the passage of charged and uncharged species across cellular membranes. It is mandatory however, to use Eq. 6-3 appropriately. The equilibrium condition $\Delta\tilde{\mu}$ = 0 can only be applied to every point of a system that acts, insofar as a given molecule or ion is concerned, as a continuous region having no material partitions and through which the molecule or ion can freely move (Fig. 6-1). In the particular case of biological membranes, we can write Eq. 6-3 for a given species provided *it goes through the membrane and the whole system is in equilibrium*.

6-2. Biological Resting Potentials

All living cells show a difference in electrical potential across their cellular membranes. This potential is called *the resting potential* and

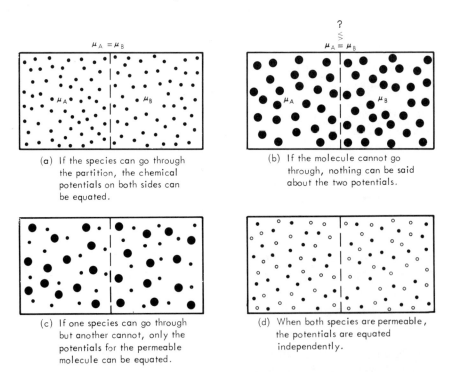

(a) If the species can go through the partition, the chemical potentials on both sides can be equated.

(b) If the molecule cannot go through, nothing can be said about the two potentials.

(c) If one species can go through but another cannot, only the potentials for the permeable molecule can be equated.

(d) When both species are permeable, the potentials are equated independently.

Fig. 6-1. Examples of the use of the equilibrium condition $\mu_1 = \mu_2$.

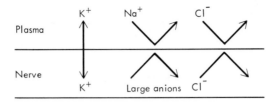

Fig. 6-2. The resting potential in nerve can be understood in physical terms by assuming that, to a first approximation, only K^+ ions can permeate the nerve.

is best understood in nerve axons, although their physical origin is similar in other cases. Before considering the quantitative part of the phenomenon, let us look at it from a qualitative point of view. Usually, the interior of the cell has a high potassium content and a low sodium content; the situation in the extracellular substance is exactly the opposite. While there are other ions that insure total electrical neutrality, both inside and outside the cell, to a first approximation, only K^+ goes through the nerve membrane in the resting state (Fig. 6-2).

Since the concentration of K^+ is larger inside than outside the cell, there will be an initial outward flow of potassium, as shown in Fig. 6-3. Although the negative ions will tend to follow K^+, they cannot go through the membrane; therefore, for every K^+ ion that leaves the axoplasm there is an unbalanced negative charge left behind; moreover, when this charge goes to the plasma-side exterior, it adds an unbalanced positive charge there. Thus, the net positive charge accumulation exerts a force on the next positive ion in a direction that tries to oppose the motion. Each successive ion that passes from the concentrated to the dilute compartment increases this force. Eventually, the electrical force exactly balances the entropic force that tends to equate the concentrations on both sides, and equilibrium is reached.

When equilibrium is attained, there is a net accumulation of positive and negative charges—outside and inside the membrane respectively, as shown in Fig. 6-3. The macroscopic manifestation of this accumulation is an electrical potential difference that can be measured with standard electrometric equipment if special precautions are taken. We can calculate the theoretical value of this potential difference by using the equilibrium condition for the chemical potential of the positive ions.

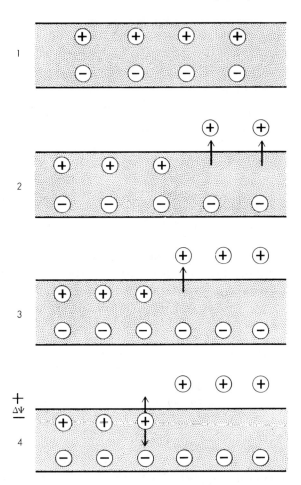

Fig. 6-3. The origin of the Nernst potential. At the start the net charge inside is zero. Positive ions can go through the membrane. They move to the outer compartment to equate concentrations. A *net* positive charge develops outside, and a *net* negative charge is left behind in the inner compartment. The resultant electrical force prevents the motion of other positive ions; the separation of charge is detected in the form of a difference in electrical potential.

6-3. Quantitative Derivation of the Electrical Potential: Nernst Equation

Since potassium ions can go through the membrane, we can write, in light of the previous discussion,

$$(6\text{-}4) \qquad \tilde{\mu}_{K^+}^{\text{inside}} = \tilde{\mu}_{K^+}^{\text{outside}} ;$$

that is, the difference in electrochemical potential for K^+ between the inside and outside of the nerve membrane is zero:

$$(6\text{-}5) \qquad 0 = \overset{\sim}{\mu}{}^{\text{in}}_{\text{K}} - \overset{\sim}{\mu}{}^{\text{out}}_{\text{K}}$$

$$= RT \, \ln \frac{(\text{K})^{\text{in}}}{(\text{K})^{\text{out}}} + ZF \, (\Psi^{\text{in}} - \Psi^{\text{out}}).$$

Recalling that z for K^+ is $+1$ and rearranging terms, we obtain the final expression for the electric potential difference:

$$(6\text{-}6) \qquad \Delta\Psi = \Psi_{\text{out}} - \Psi_{\text{in}} = \frac{RT}{F} \, \ln \frac{(\text{K})^{\text{in}}}{(\text{K})^{\text{out}}}.$$

This formula is called *Nernst's Equation*. As the concentration of K^+ in the nerve is larger than the concentration in the plasma, the ratio $K^+_{\text{in}}/K^+_{\text{out}}$ is larger than 1, so the natural logarithm is positive, as is also the potential difference ($\Psi_{\text{out}} - \Psi_{\text{in}}$). The potential outside, then, is larger than the potential inside; in other words, the inside is negative with respect to the outside, as predicted.

What are the magnitudes of the expected potentials for a given concentration ratio? The potentials for some useful concentration ratios at $37°$ C are given in Table 6-1.

TABLE 6-1. Electrical Potential Difference for Some
Concentration Ratios

CONCENTRATION RATIO (c_1/c_2)	POTENTIAL DIFFERENCE $(\Delta\Psi)$
2:1	17 mV
10:1	58 mV
100:1	116 mV

Notice that the change in potential with concentration ratio is relatively small. The composition of axoplasm (inside) and plasm (outside) of the squid *Loligo* is given for reference in Table 6-2.

TABLE 6-2. Ionic Concentrations in the Nerve of the
Squid *Loligo* (millimoles/liter)

	K^+	Na^+	Cl^-
Nerve (inside)	321	101	82
Seawater (outside)	13	498	520

Typical resting potentials measured in various neurons are given in Table 6-3.

TABLE 6-3. Resting Potentials of Some Nerve Cells (mV)

Mammalian sympathetic ganglion cells	−70
Amphibian motoneurons	−40 to −60
Electric lobe of *Torpedo*	−35 to −60
Ganglion cells of *Aplisia*	−30 to −60

Source: Adapted from John C. Eccles, *The Physiology of Nerve Cells* (Baltimore: Johns Hopkins Press, 1968).

6-4. Microscopic Separation of Charge

The concentrations that appear in the Nernst Equation are the original concentrations of the ions inside and outside the membrane. The question now arises as to whether we are allowed to use the original concentrations; after all, one could argue that both concentrations have changed in the process of building up the electrical potential difference. Actually, the changes are so small that they could never be detected by macroscopic measurements. This also explains how it is possible to "violate" the principle of conservation of charge on a microscopic scale; conservation of charge is a macroscopic observation, and as long as we cannot detect variations, it holds.

An analogous remark can be made of the once all-powerful law of conservation of mass. For example, when oxygen and hydrogen react to form water, a certain amount of mass is lost in the form of energy, according to the now well-known relativistic considerations. However, the amount of mass that disappears is only 10^{-12} (0.000000000001) kilogram per mole of water formed, and it is not possible to detect such a minute change by weighing the reacting mixture before and after the reaction occurs. Therefore, we say that from the macroscopic point of view mass is conserved in a chemical reaction.

The number of charges that have to go through the membrane in order to produce a certain biological potential is calculated using a property called the *capacitance* of the membrane. This is one of the important properties of surfaces—such as biological membranes—which are involved in electrical phenomena. The capacitance, C, of the membrane indicates its ability to keep charge separated when there is a certain voltage across it. It is defined by the ratio of the maximum charge the system can separate at a given electrical potential difference,

$$C = \frac{Q}{\Delta\Psi} ;*$$

in which Q is the charge.

One side of the surface develops a positive charge, as indicated in Fig. 6-4, and the other side acquires a negative charge of the same absolute value. It is possible to measure the capacitance of a nerve axon by using special electrical techniques, and it is found that 1 square centimeter of membrane has a capacitance of C = 0.000001 farad per square centimeter = 1 microfarad per square centimeter.

Let us now calculate what is the charge separation needed to achieve a 0.001 volt (1 millivolt) potential difference at this capacitance value:

$$Q = CV = 10^{-6} \times 10^{-3} = 10^{-9} \text{ coul/cm}^2 .$$

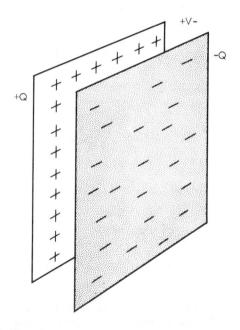

Fig. 6-4. The electrical capacitance of two parallel membranes is given by the amount of charge that can be separated per unit voltage.

*The unit of capacitance is the farad, which is the capacitance that can separate a charge of 1 coulomb across a potential difference of 1 volt.

But we said that 1 mole of a singly charged ion carries 96,500 coulombs. Then 10^{-9} coulombs are the charge in $10^{-9}/96,500$ moles, or about 10^{-14} moles per square centimeter of membrane—a very small number. This corresponds to the passage of only 6×10^{-9} molecules through each square centimeter of the membrane to develop 1 millivolt potential difference. The change in the initial ion concentration across the membrane is so small that it cannot be measured by macroscopic means.

The continuous loss of K^+ and gain of Na^+ is, however, balanced by a continuous "pumping" of these ions in the opposite direction. These pumping activities are coupled to metabolic processes and, unlike the passive Nernst equilibrium, they require a supply of energy.

6-5. Action Potentials

From the functional point of view, the most important characteristic of nerve is its irritability, the property to conduct a disturbance from one region to another. This process takes place in two steps: first, there is a local disturbance in which the potential—normally negative inside with respect to outside—reverses its polarity at a point in the nerve and becomes positive for a fraction of a second (Fig. 6-5). Secondly, this local disturbance gives rise to a propagation of the action potential along the nerve (Fig. 6-6). The coupling between the

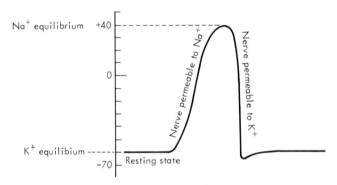

Fig. 6-5. Action potential recorded between inside and outside of giant axon in the squid. When the membrane is depolarized, it becomes permeable to Na^+. (The reason for this phenomenon is not understood.) The maximum positive value of the action potential is—to a first approximation—the Nernst equilibrium for Na^-. The membrane then becomes permeable again to K^+ and returns to the resting state.

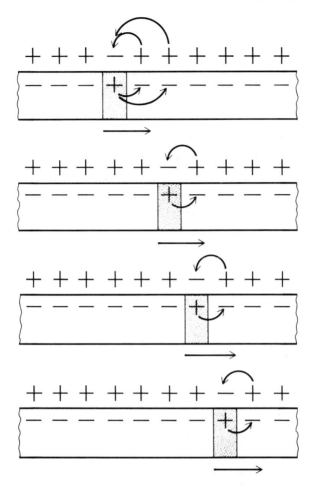

Fig. 6-6. The propagation of the nerve excitation takes place by successive depolarizations of neighboring regions. This depolarization triggers the actual permeability changes that give rise to the local event.

two processes constitutes a regenerative phenomenon, as the appearance of an action potential at a certain point depolarizes a nearby region, thus changing the permeability properties of the membrane and giving rise to an action potential at that point. In this way, the nervous impulse propagates along the nerve. There is an important difference between the two parts of this mechanism. The original disturbance is a chemical phenomenon; the propagation is an electrical phenomenon. In other words, the first step is "wet"; the second is "dry." We shall consider here only the first step, which is relevant to our discussion of the electrochemical potential equilibrium. The most widely accepted theory of the origin of the nerve impulse holds that the nerve action potential—or local disturbance—can be explained in terms of a change in the permeability properties of the axon in such a way that it becomes permeable to Na^+ and impermeable to K^+ during a fraction of a second. If that is the case, the action potential should go to the equilibrium value of Na^+ during part of the change. It does, to a finite approximation, and the maximum positive value for the action potential can be calculated using the Nernst Equation with

$$\Delta \Psi = \Psi_{in} - \Psi_{out} = \frac{RT}{F} \ln \frac{Na^{out}}{Na^{in}}$$

A lucid detailed analysis of the propagation of the nerve impulse is given in Bernard Katz, *Nerve, Muscle, and Synapse,* and the interested reader is referred to that book for further discussion.

6-6. What Came First: Potentials or Concentration Differences?

Although we have a few physicochemical ways of relating electrical potentials to the differences in concentration across cellular partitions, it is not clear why we find a potential difference across almost every cellular membrane we look at. In other words, do the potential differences have any biological significance? The answer is both yes and no. On one hand, it is hard to believe that the electrical potentials appeared before the concentration differences; if anything, the potential difference across a biological membrane applies very large forces on the membrane—although one cannot exclude the possibility that these large forces may contribute to the overall stability of the membrane. On the other hand, the differences in concentration *are* important because they are what distinguish the cellular environment from the external world. Thus, the cell maintains a favorable internal environment by performing metabolic work, and the potential difference arises as a consequence of the difference in the ionic concentration. In many cases, only one or two ions are pumped into or out of the cell, and once the electrical potential difference is set up, other ions are distributed in equilibrium according to the constraints imposed by the potential difference and the requirement that $\tilde{\mu}_{in} = \tilde{\mu}_{out}$ for that particular ion. This is the case, for example, in the frog skin in which Na^+ is actively pumped, thus creating an electrical potential difference while Cl^- is distributed according to its equilibrium requirements.

6-7. The Importance of Equilibrium Calculations; Activity Coefficients

What knowledge do we gain by figuring out whether a real biological potential follows the Nernst Equation? Given a cell, say having a concentration C_1 of ion 1 and a difference in potential between inside and outside, we can apply different concentrations of the ion outside the cell and observe whether the change in potential between the two regions obeys the Nernst Equation. If it does, we already know three things:

1. The membrane is permeable to the ion.
2. This particular ion moves passively across the membrane.
3. Equilibrium is reached during the experiment.

As an example of such a calculation, consider the unicellular alga *Nitella* which has a cellular concentration of 500 millimolar in potassium ion; let the surrounding solution be 0.5 millimolar in the same ion. The potential that would arise if K^+ were permeable and in equilibrium would be 178 millivolts—outside positive with respect to inside. This is the potential actually measured; we can conclude that K^+ can go through the membrane and that it is in equilibrium; we can also conclude that K^+ is the only ion that goes through the membrane. Suppose we now ask the same question about K^+ in the bile of a rabbit (0.5 mM) and compare it with the concentration in the serum (3.9 mM). The potential difference calculated for equilibrium is 28 millivolts, using the Nernst Equation. However, the measured potential is 16 millivolts. It looks, then, as if the ion is not in equilibrium across the membrane. Actually, it is; what has gone wrong? The main problem comes from assuming ideal behavior for the solution by writing the electrochemical potential as

$$(6\text{-}7) \qquad \Delta\mu = RT \ln \frac{C_1}{C_2} + zF\,\Delta\Psi.$$

The behavior of the concentration term—the first term of Eq. 6-7—in real solutions is slightly different, and it follows the equation

$$(6\text{-}8) \qquad \Delta\mu = RT \ln \frac{\gamma_1 C_1}{\gamma_2 C_2} + zF\,\Delta\Psi$$

in which γ_1 and γ_2 are called the *activity coefficients* of the ion in solutions 1 and 2. These coefficients are essentially a correction factor. The product γC is called the activity, a, and we can think of it as an *equivalent concentration*. The difference in electrochemical potential is then written as

$$(6\text{-}9) \qquad \Delta\tilde{\mu} = RT \ln \frac{a_1}{a_2} + zF\,\Delta\Psi.$$

6-8. Are Activities Arbitrary Quantities? The Glass Electrode

There is a way of measuring the activity coefficients, so these calculations are not as arbitrary as they may seem at first. Activity measurements depend on the existence of glass, which acts as if it were permeable to some ions but impermeable to others. Once this glass is available, it is possible to make special probes in the shape of a test tube containing the given ion at a known activity, as shown in Fig. 6-7. This glass electrode, or probe, is dipped into the solution whose activity we wish to measure and the difference in electrical potential between the electrode and test solution is recorded. Since the special glass permits the passage of the ion of interest, we can write the equilibrium condition for this ion as

$$\tilde{\mu}_{ion}^{electrode} = \tilde{\mu}_{ion}^{solution}.$$

When the explicit forms of the electrochemical potentials are introduced, we obtain

$$RT \ln \frac{a_{ion}^{solution}}{a_{ion}^{electrode}} = zF\,(\Psi_{electrode} - \Psi_{solution}).$$

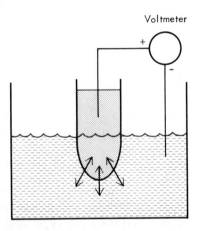

Voltmeter

Fig. 6-7. The glass electrode is essentially a membrane permeable to a single ion; the potential difference across is, then, a Nerst potential.

As the activity of the ion in the electrode is known and the electrical potential difference can be measured, we can find the activity of the ion in solution from the Nernst Equation. Although K^+, Na^+, and other glass electrodes are available in the market, the hydrogen electrode (Prob. 6-7) is the one most used in the laboratory because it gives the pH of the solution. Recently, it has been possible to make small glass microelectrodes that can go into a kidney tubule measuring the activities of different ions *in situ* (Fig. 6-8).

6-9. Gibbs-Donnan Equilibrium

We shall now consider another form of biophysical equilibrium, Donnan equilibrium, which is characteristic of solutions containing both small ions and large charged molecules. To fix ideas, let KCl be in two compartments at two different concentrations, c_1 and c_2, and let a large charged molecule, such as a protein, be present in the second compartment. This case is typical of cells, where polyelectrolytes (large molecules with many charges) are always present. If the protein inside has ν negative charges per molecule, the positive ions (K^+) will be attracted toward the compartment where the protein is, whereas Cl^- ions will be expelled, as shown in Fig. 6-9. Eventually, a potential difference will appear at equilibrium with the side where the protein is having the positive sign. Since the macroscopic principle of conservation of charge must be obeyed, the total positive charges must equal the total negative charges in each compartment,

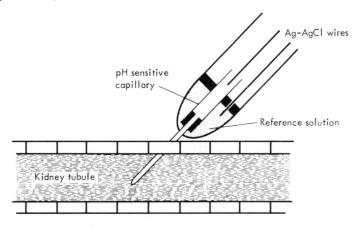

Fig. 6-8. Schematic drawing of a fine-tip pH microelectrode used to measure $[H^+]$ in kidney tubules.

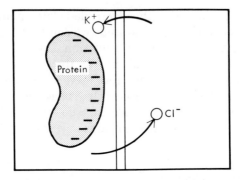

Fig. 6-9. Donnan equilibrium. Since both ions can go through the membrane, the potential difference would be zero at equilibrium in the absence of a charged protein. The large polyelectrolyte causes a redistribution of ions and gives rise to an electrical potential difference.

(6-10) $$[K^+]_1 = [Cl^-]_1 \quad \text{and}$$

(6-11) $$[K^+]_2 = [Cl^-]_2 + [P]\,v,$$

in which the bracketed quantities are concentrations, expressed in moles per liter, and P is the protein. We define the Donnan ratio as

$$r = \frac{[K_1^+]}{[K_2^+]}.$$

The Nernst Equation applies at equilibrium to both K^+ and Cl^-; thus,

(6-12) $$\Psi_2 - \Psi_1 = \frac{RT}{F} \ln \frac{K_1}{K_2} \quad \text{and}$$

(6-13) $$\Psi_2 - \Psi_1 = -\frac{RT}{F} \ln \frac{Cl_1}{Cl_2} = \frac{RT}{F} \ln \frac{Cl_2}{Cl_1}$$

Then, equating 6-12 and 6-13, we obtain

(6-14) $$\frac{Cl_2}{Cl_1} = \frac{K_1}{K_2} = r.$$

We can rewrite Eq. 6-14 as

(6-15) $$Cl_2\,K_2 = K_1\,Cl_1.$$

Therefore, the products of the concentration of the ions are equal on both sides when Donnan equilibrium is obeyed. Introducing Eq. 6.14 into Eqs. 6.10 and 6.11, we obtain an explicit expression for r in terms of the protein concentration, the charge in the protein, and the concentration outside the protein compartment:

$$(6\text{-}16) \qquad r = \frac{v[P]}{2[K_i^+]} + \sqrt{\left(\frac{vP}{2K_i^+}\right)^2 + 1}.$$

If the inequality $\qquad v[P] \ll 2[K_i^+]$

holds, we obtain the limiting formula

$$(6\text{-}17) \qquad r = 1 + \frac{v[P]}{2[K_i^+]}.$$

6-10. Gibbs-Donnan Equilibrium in the Erythrocyte

In the erythrocyte, the Gibbs-Donnan equilibrium is caused by the impermeability of the membrane to hemoglobin, the effect of other proteins and ions being negligible. Sodium and potassium, for example, do not contribute to the Donnan effect because

1. The membrane is not permeable to these ions, so they do not appear in the Donnan ratio.
2. Their effects exactly counterbalance; K^+ has a larger concentration in the cell whereas Na^+ is concentrated in the plasma (Table 6-4).

TABLE 6-4. Concentrations of Some Solutes in the Erythrocyte and in Blood Plasma (millimoles/liter)

	K^+	Na^+	Ca^{++}	Cl^-	Glucose
Plasma	5.35	144	3.2	111	4.3
Cells	150	12-30	–	73.5	4.3

The relevant ions are, in the present context, chloride, bicarbonate, and hydroxil. The theoretical equalities relating their individual concentrations in cells and plasma are:

$$\frac{(Cl^-)\ cell}{(Cl^-)\ plasma} = \frac{(HCO_3^-)\ cell}{(HCO_3^-)\ plasma} = \frac{(OH^-)\ cell}{(OH^-)\ plasma}$$

The actual (experimental) values of these three ratios are 0.60, 0.685, and 0.63 for the chloride, bicarbonate, and hydroxil distributions, respectively.

6-11. Chemical Potential and Osmotic Equilibrium

Consider a system subdivided into two parts, A and B, both containing two *uncharged* components, 1 and 2 (Fig. 6-10). Furthermore, assume that the partition is permeable to one of the components, say 1, and impermeable to the other. At the partition, we can write for $\Delta \tilde{\mu}$ when the system has reached equilibrium, $\mu_1^{(A)} = \mu_1^{(B)}$. Writing this equilibrium condition in its explicit form it becomes

$$(6\text{-}18) \qquad RT \ln \frac{C_1^A}{C_1^B} + \overline{V}_1 (P^A - P^B) - \overline{S}(T^A - T^B) = 0$$

at the membrane. If the pressure and temperature across the membrane are the same on both sides and if the A molecules are uncharged, the previous equation reduces to

$$(6\text{-}19) \qquad RT \ln \frac{C_1^A}{C_1^B} = 0$$

which yields the equilibrium condition $C_1^A = C_1^B$. Then, *in the absence of any pressure or temperature differences, the concentration of the permeable component would be the same on both sides.* On the other hand, if we allow for a pressure difference across the partition at equilibrium, Eq. 6-18 gives

$$RT \ln \frac{C_1^A}{C_1^B} + \overline{V}_1 (P_A - P_B) = 0,$$

or, rearranging terms,

$$(6\text{-}20) \qquad \frac{RT}{\overline{V}_1} \ln \frac{C_1^A}{C_1^B} = (P_B - P_A).$$

The pressure difference $(P_B - P_A)$ is called the osmotic pressure, and

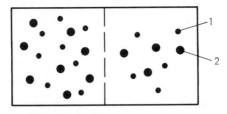

Fig. 6-10. A two-component, two-compartment system in which only one species can go through the partition.

it is denoted by $\Delta\pi$. When the concentration of component 2 is small on both sides, it can be shown that Eq. 6-20 reduces to

$$RT\ \Delta C = \Delta\pi \qquad \text{(Van Hoff's formula)}$$

in which

$$\Delta C = C_2^B - C_2^A$$

and

$$\Delta\pi = P_B - P_A .$$

It is interesting to note that the concentrations of component 2 on both sides have appeared in the final result even though the equilibrium condition was never given in terms of component 2. Species 1 serves as a "messenger," informing component 2 that the two sides are connected.

6-12. Physical Meaning and Biological Implications of Osmotic Pressure

Let us consider a semipermeable membrane, which is permeable to A but not B, mounted on a piston that can move in the x direction (Fig. 6-11). The piston naturally divides the total system into two regions, A and B, whose relative volumes can be changed by moving the piston to right or left, although the total volume of the system does not change. It is clear that whenever the piston moves from right to left permeable molecules will be forced to pass from left to right (in most real cases the permeable molecule is an incompressible

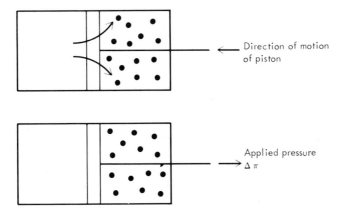

Fig. 6-11. The piston will move to the left unless a pressure $\Delta P = \Delta\pi$ is applied to the right. This pressure increases the hydrostatic pressure of water in the right compartment and prevents further water motion.

liquid like water). Conversely, if the permeable molecules pass from left to right, the relative volumes of left and right compartments will change and the piston will be forced to move from right to left. Assume that at "time zero," when the experiment starts, the concentration of the permeable molecules in the left is larger than their concentration in the right. Following the laws of thermodynamics, they will flow in such a direction as to equate the two concentrations; that is, they will move from the left compartment to the right compartment until the two concentrations are equal. As we have seen, the final pressure difference in this case is zero if the process takes place at constant temperature. Suppose now that the system starts as before, but instead of allowing the piston to move we hold it in its original position. Obviously, a force must be applied to prevent the motion of the cylinder. The force per unit area applied on the cylinder is the osmotic pressure previously calculated. We now consider a balloonlike permeable membrane, rather than a piston, and instead of talking about right and left sides, we shall refer to an inside and an outside, and instead of saying that membrane moves to left or right we shall say that it expands or that it contracts. But the problem is essentially the same. Take a soap bubble, for example, and place it in an atmosphere of CO_2 to which the bubble is permeable. Since the concentration of CO_2 outside is larger than inside (we assume the bubble contains air in its interior), CO_2 enters into the bubble, thus increasing its volume. If the initial external concentration of CO_2 is not very large, a small inflow of the gas will suffice to equate the two concentrations; in this case, the bubble will slightly increase in volume and an equilibrium state will eventually be reached. As in the case of the piston considered above, there will be a force at equilibrium; this is the surface force that holds the soap bubble together. Common experience tells us that after a certain volume is exceeded the bubble explodes; this happens because the resultant surface forces are larger than the forces holding the membrane together.* When the complementary experiment is performed and CO_2 is introduced inside while the external atmosphere is air, the bubble contracts.

Biological membranes behave like this membrane. Water is the main permeable molecule; salts and other solutes are usually impermeant. Animal cells possess a typical "unit membrane," which is composed of a bi-lipid protein layer; this membrane is reasonably

---

*An illuminating account of these forces is given in *Soap Bubbles and the Forces Which Mold Them*, Charles V. Boys' scientific classic (New York: Doubleday and Co., Inc., 1959).

elastic but it cannot sustain excessive surface forces. For this reason, a large difference in concentration may cause either the irreversible rupture of the membrane or the complete loss of water from the cell when the concentration of impermeable solutes in the medium is much smaller or much larger than in the cell, respectively. This is the reason why survival manuals recommend not drinking seawater; its entry into the organism in large quantities causes irreparable water loss.

6-13. Turgor Pressure and Stomatal Opening

Unlike animal cells, plant cells have a rigid wall which, though permeable to all molecules, prevents excessive volume increase in the inner unit membrane (which is identical, or at least very similar, to the animal membrane). As a result, a plant cell placed in a hypo-osmotic (low salt concentration) medium becomes turgid,* but it maintains its integrity (Fig. 6-12). Thus, pressure gradients in plants can reach very large values—up to 150 atmospheres.

An important mechanism in plant physiology—which is directly related to the existence of osmotic gradients—is the control of stomatal size. The stomata, or openings in the leaf epidermis, are the passages that regulate the exchange of gases taking part in photosynthesis (Fig. 6-13).

We shall see in the next chapter that whenever a substance moves between two regions separated by a partition, the motion (diffusion) is proportional to the area of the partition. In the case of diffusion across openings such as stomata, however, diffusion is a much more efficient process and the rate at which gases move is about one hundred times the value it would have if stomata were covered by a membrane (Fig. 6-14). Thus, the structural problem of keeping the leaf as an integral structure while permitting the quick exchange of gases through an area equivalent to that of the whole leaf has been solved in plants by the development of small openings. The problem presented by stomata, however, is the need to open and close them according to photosynthetic needs. The mechanism is based on changes in the turgor pressure of the guard cells, which are the cells lining the stomatal opening. These cells possess two walls, an inner wall and an outer wall, of different thicknesses (see Fig. 6-15). When water rushes into the guard cells, the expansion of the two walls is uneven, as the outer wall can expand more easily than the inner wall,

*This hardening is easily seen when a peeled potato is placed in water.

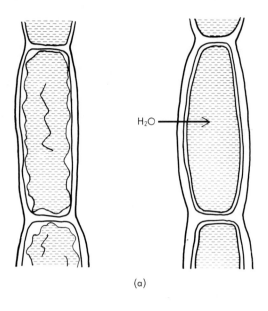

(a)

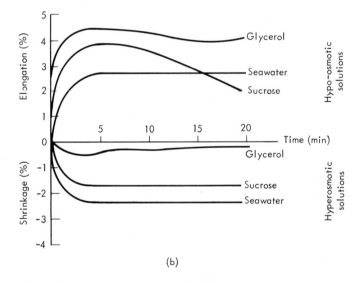

(b)

Fig. 6-12. Unlike animal cells, plant cells have a thick cellular wall that prevents the membrane from expanding to the point of disintegration when placed in distilled water (*a*). A series of typical osmotic experiments on algal cells is shown in (*b*): Cells either elongate or shrink, depending on whether the environment is hypo- or hyperosmotic. (F. Gessner and N. Schramm, in *Marine Ecology*, ed. by O. Kinne (London: Wiley-Interscience, 1972).

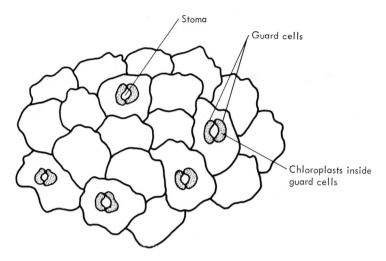

Fig. 6-13. Typical distribution of stomata on the surface of a dicotyledonous leaf.

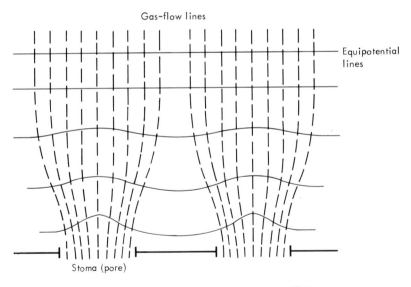

Fig. 6-14. Gas flow through stomata is a very efficient process; stomatal openings provide a "lens" effect that enhances flow.

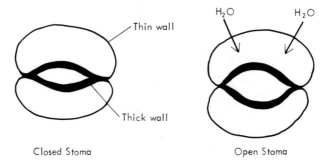

Fig. 6-15. Stomata open when H_2O enters the guard cells because the thicker walls lining the opening cannot expand.

thus leading to stomatal opening. Why does water move from the surrounding epidermal cells into the guard cells? Guard cells are the only cells in the epidermis that possess chlorophyll. They can carry out photosynthesis and decrease or increase solute concentration relative to the neighboring cells. The general changes in guard cells which lead to stomatal opening and closing are given in Fig. 6-16, although the detailed mechanism is not yet fully understood at the molecular level.

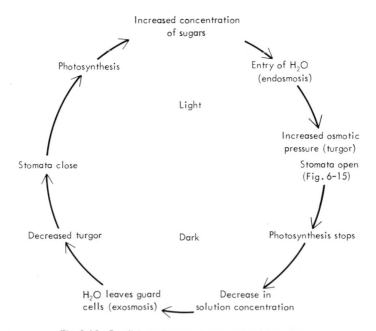

Fig. 6-16. Possible events leading to stomatal opening.

6-14. Osmosis Is a Very Strong Force

At $0°$ C, a 0.1 molar solution of sucrose produces a pressure difference 2.5 times that of the atmosphere at sea level, while at $25°$ C (room temperature) a 1 molar solution gives a pressure 27 times larger than that of the atmosphere! Osmotic forces are probably the most important in the transport of water in most organisms. Clearly, proper consideration of osmotic pressures is fundamental in the maintenance of tissues both in vivo and in vitro. When a biological experiment requires the removal of a tissue and its handling outside of its natural environment, appropriate physiological solutions must be made in order to maintain the tissue's life. The main factors to consider when making such solutions (salines) are:

- Presence of the same chemical species—and in the same concentration—in which they are naturally bathing the cells, especially the essential ions—Na^+, K^+, HCO_3
- Proper acidity (pH)
- Adequate supply of oxygen
- Adequate supply of metabolites
- Correct osmotic pressure

In terms of keeping the osmotic pressure within normal levels, the problem arises because there are many molecules in solution, some of which are charged. Since osmotic pressure depends on the number of particles present in solution, the total osmotic pressure in the presence of concentration gradients in different nonelectrolytes (1, 2, 3, ... n) is given by the sum of their individual osmotic pressures:

$$\Delta\pi_{total} = \Delta\pi_1 + \Delta\pi_2 + \ldots + \Delta\pi_n$$
$$= RT (\Delta C_1 + \Delta C_2 + \ldots + \Delta C_n).$$

To a first approximation, we can calculate the osmotic pressure developed by gradients of electrolytes in the same way, and assuming that the electrolyte dissociates completely into its constituents ions (strong electrolyte). For example, if we had a difference in concentration of 1 molar di-sodium sulfate, Na_2SO_4, across a membrane impermeable to Na_2SO_4, we could write the total Van Hoff's expression as if it were caused by 2 moles of Na^+ and 1 mole of SO_4; that is,

$$\Delta\pi = RT (\Delta C_{NA^+} + \Delta C_{SO_4})$$

in which
$$\Delta C_{NA^+} = 2M$$

and
$$\Delta_{SO_4} = 1M$$

This expression turns out to be incorrect, however, because the non-ideal interactions among ions reduce the effective number of available particles, so the pressure is smaller than what the ideal formula gives. We can introduce a correction factor, Φ, the Van Hoff's coefficient, so that the osmotic pressure becomes

$$\Delta\pi = RT\,\Phi\,(\Delta C_1 + \Delta C_2 + \ldots + \Delta C_n)$$

for Na_2SO_4 at $25°$ C and 0.1 molar equals 0.793, while for sucrose, under the same conditions, it is 1.008, or almost ideal. We include a table with values of osmotic pressures for different solutes at 25 degrees Centigrade (Table 6-5).

TABLE 6-5. Osmotic Pressure (atm)

MOLARITY	$CaCl_2$	NaCl	Glucose
0.01	0.7	0.5	2.4
0.20	12.3	8.9	7.2
	18.5	13.2	9.6
0.40	25.3	17.6	12.1
0.50	32.2	21.9	14.5
0.70	47.2	30.5	19.4
1.00	75.0	43.6	

6-15. Colloid Osmotic Pressure and the Motion of Water in Capillaries

In systems that obey the Gibbs-Donnan equilibrium, the redistribution of ions produces a difference in osmotic pressure given (to a first approximation) by the sum total of the concentration differences:

$$\Delta\pi = \Delta[P] + \Delta[C_1] + \Delta[C_2] + \ldots + \Delta[C_n],$$

in which $\Delta[P]$ is the difference in protein concentration across the membrane and $\Delta[C_1]$ through $\Delta[C_n]$ the difference in concentration for all other species. As a result, water tends to move into the compartment containing the charged protein. This particular example of osmotic pressure is usually called *colloid osmotic pressure*. Ernest Starling hypothesized—and Eugene Landis showed experimentally—that this pressure is responsible for the entry of water into the blood capillaries at their venous end and its exit at the arterial end (Fig. 6-17). The colloid pressure in the capillaries can be calculated from the concentrations of permeable and impermeant solutes (Table 6-6); this comes out to 36 centimeters of H_2O.

TABLE 6-6. Concentrations of Solutes in Lymph
and Plasma

	PLASMA	LYMPH
Protein (%)	6.85	2.61
Sugar (mg/100 ml)	123	124
Calcium (mg/100 ml)	10.4	9.2
Chloride (mg/100 ml)	392	413

The hydrostatic pressure of the blood—caused by the pumping of the heart—is 44 centimeters of H_2O at the arterial end and 17 centimeters of H_2O at the venous end. Thus, the net pressures are 8 centimeters of H_2O at the arterial end (outward) and 19 centimeters of H_2O at the venous end (inward). The individual contributions and net effects are indicated in Fig. 6-18.

6-16. Reflection Coefficients

The osmotic examples discussed so far have included only membranes that act like ideal molecular sieves: Some molecules pass through freely while others are completely excluded from crossing the boundary. Water and very large proteins are characteristic of this extreme type of behavior, but most uncharged water solutes of low molecular weight are somewhere in between. As a result, the membrane does not act as an ideal sieve; it allows a certain fraction of the given solute to go through the membrane and rejects the rest. Since the only molecules that contribute to the osmotic pressure are those *excluded* from the membrane, only a fraction of the total solute molecules are osmotically active. The osmotic pressure expression then changes to

(6-21) $$\Delta\mu = \sigma RT\ \Delta c,$$

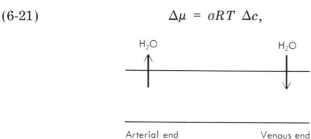

Fig. 6-17. Water enters blood capillaries at the venous end and leaves at the arterial end.

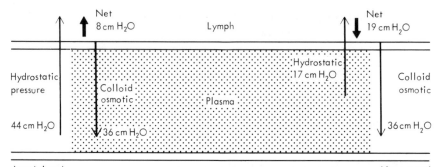

Fig. 6-18. The *net* pressure is the difference between the heart's hydrostatic pressure and the colloid osmotic pressure.

in which the reflection coefficient gives the fraction of solute molecules that are "reflected" back and therefore contribute to the osmotic pressure.

Clearly, the maximum value σ can have is 1, if all the solute molecules are "reflected"; and the minimum value is 0, if all the molecules go through the membrane (Fig. 6-19).

For all other cases, the reflection coefficient will be between 0 and 1. Therefore, the reflection coefficient, σ, will give an indication of how permeable the membrane is to a given solute. Reflection coefficients for some biological and nonbiological membranes are given in Table 6-7.

TABLE 6-7

MEMBRANE	SOLUTE	REFLECTION COEFFICIENT
Toad skin	Acetamide	0.89
	Thiourea	0.98
Nitella translucens	Methanol	0.5
	Ethanol	0.44
	Urea	1.00
Human red blood cells	Urea	0.62
	Ethelyne glycol	0.63
	Melonamide	0.83
Visking dialysis tubing	Urea	0.013
	Glucose	0.123
	Sucrose	0.163

Source: A. Katchalsky, and P. Curran, *Non Equilibrium Thermodynamics in Biophysics* (Cambridge, Mass.: Harvard University Press, 1965).

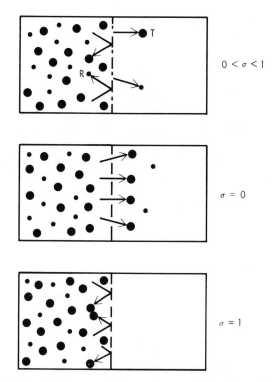

Fig. 6-19. The reflection coefficient gives an indication of the number of molecules that are reflected back into the same compartment (R) compared to the number that are transmitted (T). When the reflection coefficient is 1, all the molecules are reflected; when it is 0, all the molecules are transmitted. The reflection coefficient must be specified for each molecule and each membrane, since the relative sizes of the membrane passages and solute size will determine the number of molecules transmitted compared with the number of molecules reflected.

6-17. A Possible Way to Measure σ

The pressure difference that must be applied to stop water flow is the observed osmotic pressure $\Delta\Pi$. On the other hand, the ideal osmotic pressure for small uncharged solutes to which the membrane is completely impermeant is

$$\Delta\pi_{ideal} = RT\,\Delta c.$$

From Eq. 6-21, it follows that

$$\sigma = \frac{\Delta\pi}{\Delta\pi_{ideal}};$$

that is, σ is the ratio of the observed to the ideal osmotic pressure, $RT \Delta C$.

6-18. Osmotic Pressure in the Presence of Several Solutes

As a given membrane may have different sigma's for different molecules, whenever there exists a concentration difference in several species it is necessary to add their individual osmotic pressures, including the appropriate reflection coefficients. Thus, the total osmotic pressure set up across a membrane by two substances A and B at concentrations ΔC_A and ΔC_B is

$$\Delta \pi_{\text{total}} = \Delta \pi_A + \Delta \pi_B ,$$

or
$$\Delta \pi_{\text{total}} = \sigma_A RT \Delta C_A + \sigma_B RT \Delta C_B$$

in which σ_A and σ_B are the reflection coefficients for A and B, respectively.

6-19. Osmotic Regulation and Habitat

Osmotic properties are one of the important evolutionary forces that dictate the physiological equipment an animal must possess if it is going to survive in a given environment. Depending on the specific habitat in which they live, animals have devised various methods for maintaining the constancy of the internal environment. Prosser et al. point out in their book on *Comparative Animal Physiology* that the differences are not necessarily qualitative, as animals with a similar form of osmotic control may not be able to survive in the same type of osmotic environment. It turns out that *quantitative* differences in osmotic mechanisms make it possible for many species to live in a typical environment.

Some of the normal regulatory mechanisms are listed in Table 6-8. In *marine animals*, for example, a method of salt secretion is absolutely vital. In *freshwater* habitats, on the other hand, salt must be retained and water eliminated, as depicted in Fig. 6-20. In air-land animals body coverings prevent loss of water. An optimum method of retaining water while excreting large amounts of salt is achieved by the mammalian kidney which excretes concentrated urea.

TABLE 6-8

OSMOTIC CHARACTERISTICS	PRINCIPAL MECHANISMS	EXAMPLES
Osmotic adjustment	No volume regulation	Marine invertebrates' eggs
	Volume regulation	Marine molluscs
Fair osmoregulation in hypotonic media	Selective absorption of salts from medium, kidney reabsorption or secretion, low permeability	Carcinus (crab)
Unlimited regulation in hypotonic media	Hypotonic copious urine, salt reabsorption or water secretion, low surface permeability	Crayfish Amphibia
	Water impermeability	Freshwater embryos
Maintenance of hypertonicity in all media	Urea retention	Elasmobranchs
Regulation in moist air	Low skin permeability, salt absorption from medium, salt reabsorption in kidney	Frog Earthworm
Regulation in dry air	Impermeable cuticle, hypertonic urine	Insects
	Hypertonic urine, water reabsorption in kidney	Birds and mammals

Source: C. L. Prosser et al., *Comparative Animal Physiology*. (Philadelphia: W. B. Saunders, 1950.)

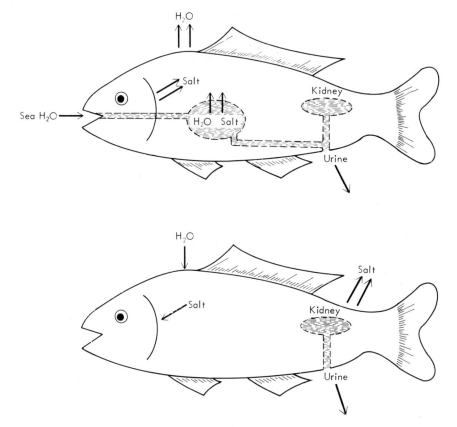

Fig. 6-20. Freshwater fish and seawater fish live in completely different osmotic environments. As a result, seawater fish conserve H_2O and excrete salt, whereas freshwater fish conserve salt and excrete H_2O. (K. Schmidt-Nielsen, *Animal Physiology*. Englewood Cliffs, N.J.: Prentice-Hall, Inc., 1961).

Problems

6-1. Derive Eq. 6-16 giving the Donnan ratio r,

$$ r = \frac{v[P]}{2[K_1^+]} + \sqrt{\left(\frac{vP}{2K_1^+}\right)^2 + 1}. $$

Hint: Start with Eq. 6-11 and introduce Eq. 6-10.

6-2. Answer the following questions qualitatively:

(a) What is r, the Donnan ratio, if the protein is uncharged?

(b) What is r, if the concentration of salt outside the protein compartment is very large?

(c) What are the electrical potential differences measured in (a) and (b) above?

6-3. The net charge on a certain protein is +8 at a specific pH. A 10×10^{-3} M solution of this protein is placed in a dyalisis bag—which is impermeable to the protein but allows water and small molecules to go through—and the dyalisis bag is placed in a beaker containing 300 mM KCl. The volume of the dyalisis bag is assumed to be constant and equal to the volume of the KCl in solution.

Answer the following questions:

(a) What is the final concentration of K^+ in the protein compartment at equilibrium?

(b) What is the osmotic pressure at equilibrium at $27°$ C if the protein molecules and ions behave ideally?

(c) What is the electrical potential difference at equilibrium? Assume that $T = 27°$ C.

6-4. Given the concentrations of K^+, Na^+ and Cl^- ions in the axoplasm (nerve) of *Loligo* and seawater (Table 6-2) given the equilibrium potentials of Na^+, K^+ and Cl^- with appropriate signs. Relate to the events that take place at the resting and action potentials. Let $T = 25°$ C.

6-5. How many K^+ ions are lost from the nerve per cm^2 to give the resting potential calculated above? What is the minimum amount of work needed to pump these ions back into the nerve? What is the minimum number of ATP molecules required to effect this work?

6-6. One of the most important indicators of the chemical environment (both organic and inorganic) is the element hydrogen, because its concentration directly reflects how acid or basic a medium is. The quantity usually employed to refer to hydrogen concentrations is the pH, defined by the decimal logarithm of the inverse hydrogen concentration, that is,

$$pH = \log_{10} \frac{1}{[H]} = - \log_{10} [H].$$

The pH is then related to the activity of hydrogen a_H, by the equation, pH = $-2.03 \ln (a_H)$. Explain how a "membrane" that is permeable only to H^+ could be used to measure pH. This is the principle of the pH electrode.

6-7. What kinds of experiments would you perform to test whether K^+ is in equilibrium across the nerve membrane if there were no way to introduce an electrode *inside* the axoplasm?

6-8. The following table gives the approximate concentrations for K^+, Na^+, and Cl^- in cat motoneurons:

	OUTSIDE (mM)	INSIDE (mM)
Na	150	15
K	5.5	150
Cl	125	9

(a) Calculate the Nernst potential for each of these ions at $38°$ C, giving appropriate signs.

(b) Given that the measured potential for a normal motoneuron is -70 mVolts (inside relative to outside), indicate which ion is in equilibrium and what ionic motions should be expected.

7

CHEMICAL POTENTIAL
IN ACTION: DIFFUSION

We have so far analyzed thermodynamic problems that arise at equilibrium, when no measurement performed on the system indicates the passage of time. In that case there is no change in the value of any function of state.

If the system starts *away* from equilibrium, however, there will be a force that will drive it *toward* equilibrium, passing through a series of dynamic states. In particular, biological systems never reach equilibrium but remain in a steady state in which equilibrium is always approached but never reached. In order to discuss these nonequilibrium steady states, we must first find adequate ways of describing the forces that attempt to bring the system to equilibrium as well as the resultant motions.

7-1. The Meaning of the Chemical Potential
Away from Equilibrium

In the equilibrium situation considered for the transfer of mass across a membrane, the change in free energy was zero, and, as a result, we concluded that the potentials on both compartments are equal and no work need be done to transfer mass across the membrane. If the two chemical potentials are not the same, work will be needed to move matter from the low to the high potential.

We showed in Chapter 3 that the work done when n moles are transferred from the low to the high chemical potential is given by

(7-1) $$W = -n\ \Delta\mu.$$

Furthermore, we know that work is defined as

(7-2) $$W = fd$$

when either the force is constant over the distance d or the distance is small. Consider the latter instance; in that case, we can equate Eq. 7-1 with Eq. 7-2 to obtain

$$Fd = -n\ \Delta\mu$$

which can also be rewritten as

(7-3) $$\frac{F}{n} = -\frac{\Delta\mu}{d}.$$

The force acting on a mole of substance moving across two neighboring regions held at different chemical potentials is, then, directly proportional to the potential difference and inversely proportional to the distance; that is, the larger the chemical potential difference the larger the force, and the smaller the distance the smaller the force.*

A similar case holds for the motion of charge; the work needed to transfer a positive charge from a region at low potential to a region at high potential is

$$W = -Q\ \Delta\Psi.$$

And, again over a short distance d,

$$W = Fd.$$

Equating the two expressions for electrical work we obtain

(7-4) $$\frac{F}{Q} = -\frac{\Delta\Psi}{d},$$

which states that the force required per unit charge is the potential increment per unit distance. Clearly, if the system is left to come to equilibrium by itself, matter or charge will move from the high to the low potential and will do work on the environment.

In both cases considered above, we see that the force is a change or disturbance in an intensive property (chemical or electrical

---

*The negative sign indicates that the direction of the force is opposite to the direction of increasing potential.

potential) while the resultant motion is a variation in an extensive property (mass or charge, respectively).

We should point out that forces can only be found from potential changes in systems in which energy is conserved (see Appendix).

7-2. Visualization of the Potential

Since an extensive quantity (charge, mass, etc.) moves from the high to the low potential, the potential behaves like a height. But the situation is a little more complicated because it is not only the difference in "height," or potential, that counts but also the length of the distance over which the potential change takes place; if the change takes place over a short distance, the force will be larger than if it takes place over a long distance (see Eqs. 7-3 and 7-4).

We could represent the situation by an imaginary surface such as the one shown in Fig. 7-1 in which a ball is rolling; the force driving the ball will be larger on the steeper side of the "hill," because on this side, the same change in potential "height" will be obtained for a shorter distance.

7-3. Steady Velocity

According to Newton's first law of motion, an object on which no force is acting will either remain still or, if it is moving, it will keep on moving at constant velocity. Intuitively, it is hard to visualize this

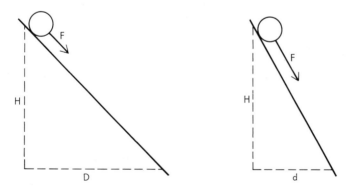

Fig. 7-1. The steeper hill—the one that has a larger increase in height per unit distance—gives a larger force on a rolling ball.

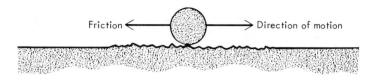

Friction ←————————→ Direction of motion

Fig. 7-2. The force of friction is opposite to the direction of motion and tends to slow down a moving particle. If the frictional force is exactly equal to the driving force, the object reaches a constant velocity; this occurs when the magnitude of the frictional force is proportional to the velocity of the moving mass.

situation because we rarely have a chance to observe motion of an object on which no frictional forces are applied.

There are other ways for an object or molecule to reach a constant velocity even when external forces act on it. As an example, consider the motion of an object acted upon by a constant force and a frictional force opposite to the direction of motion and proportional to the velocity (Fig. 7-2).

Newton's second law of motion requires that

$$\text{acceleration} = \frac{\text{sum of all forces}}{\text{mass}} .$$

Since there are two forces acting on the mass M, the driving force F, and the frictional force F_{friction}, we can write

$$\text{acceleration} = \frac{F + F_{\text{friction}}}{M} .$$

After a long time, the acceleration dies out, so that

$$0 = \frac{F + F_{\text{friction}}}{M} \quad \text{or,}$$

(7-5)
$$F = -F_{friction};$$

that is, the frictional and driving forces are equal and act in opposite directions, so there is no *net* force on the object. Since the frictional force is by assumption

$$F_{\text{friction}} = k(-v);$$

in which k is the coefficient of friction and the negative sign indicates that the force goes against the direction of motion, we can rewrite Eq. 7-5 as

(7-6)
$$F = -k\,(-v)$$

$$= +kv.$$

The final constant velocity reached is, then,

$$(7\text{-}7) \qquad\qquad v_{\text{final}} = \frac{F}{k}$$

which can be conveniently rewritten as

$$(7\text{-}8) \qquad\qquad v_{\text{final}} = \omega F$$

by defining the *mobility* ω by $\omega = 1/k$. Clearly, the mobility is the (steady) velocity an object acquires per unit driving force applied. The larger the mobility, the smaller the friction. The mobility will in general depend on the shape of the molecule, composition of the environment, etc.

7-4. Explicit Expression for Steady Flow

We can now utilize kinetic reasoning to give an expression for the steady flow of a collection of molecules at uniform concentration c and which move under the influence of some constant force F.

Refer to Fig. 7-3. The number of molecules of the substance which pass through an area A in a time t is given by all the molecules that are close enough to the left of the area A to make it through in that time. These are, as in Sec. 4-7, the molecules within a distance

$$d = vt.$$

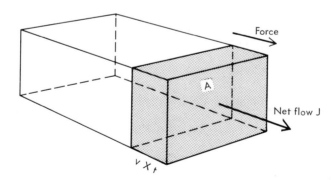

Fig. 7-3. Diagram used in the derivation of a steady-flow expression. If the average net velocity of the molecules is v, it takes a time d/v for a molecule to move a distance d; conversely, in time t only molecules that are within $v \times t$ of some cross section of the container will reach this area. The shaded volume in the figure corresponds to the molecules which in time t will cross the surface at the right.

The total number of moles in the volume tvA is

$$n = cvtA,$$

in which c is the concentration of solute in this volume. The number of moles that go through in unit time is

$$J = \frac{n}{t} = cvA.$$

Finally, we can relate the steady velocity to the driving force and mobility (Eq. 7-8) to obtain

(7-9) $$J = cF\omega A.$$

7-5. Fick's Law

The result obtained in the previous section can be restated by defining a force in terms of the spatial variation in chemical potential (or electrochemical potential). Thus, the force that appears in Eq. 7-9 is given by

$$F = -\frac{\Delta\mu}{d},$$

as discussed in Sec. 7-1. To fix ideas, consider the flow J that goes between two compartments separated by a short distance d (Fig. 7-4). The concentration c is, as before, the concentration carried by the flow; that is, the concentration *at* the thin region of thickness d.

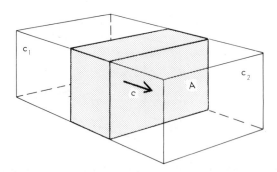

Fig. 7-4. When we calculate the diffusional flow between two regions at different concentrations, say c_1 and c_2, we may assume to a first approximation that the flow in a thin partition such as a membrane takes place in a region of constant concentration, c.

The equation for the flow J can then be written as

$$(7\text{-}10) \qquad J = - \omega c \times \frac{\Delta\mu}{d} \times A,$$

and we can make use of the fact that the concentration and chemical potential are related by a logarithm to write (Sec. 3-13)

$$\Delta\mu = RT \frac{\Delta c}{c}.$$

Introducing this equation in the flow expression Eq. 7.11, we obtain

$$(7\text{-}11) \qquad J = - \frac{\omega c}{d} RT \frac{\Delta c}{c} A$$

which, after cancellation of the concentrations in numerator and denominator, yields Fick's Law:

$$(7\text{-}12) \qquad J = - A \frac{\omega RT}{d} \Delta c.$$

This expression holds exactly only for very small concentration differences. In practice, however, it holds over large ranges; for that reason, it is best to consider it as a result of an experimental observation. The expression derived, however, shows that this observation has a reasonable thermodynamic basis.

7-6. Physical Meaning of Fick's Law

Fick's Law, Eq. 7-12, states that the steady flow of solute between two regions held at different concentrations is proportional to the area of the partition (the larger the area, the larger the flow), inversely proportional to the thickness of the membrane (the thicker the partition, the smaller the flow) and proportional to ωRT. This last quantity is usually denoted by D, the diffusion "constant," although this is a misnomer: there is nothing "constant" about the diffusion constant; it is a variable. Introducing D in 7-12, we obtain

$$(7\text{-}13) \qquad J = - \frac{A}{d} D \Delta c.$$

The meaning of the negative sign is discussed in Prob. 7-4. Table 7-1 gives typical values of D for some macromolecules of biological interest.

TABLE 7-1. Diffusion Coefficients of Some Macromolecules in Pure Water at $20°$ C

	MOLECULAR WEIGHT	D_{20}
Hemoglobin	68,000	6.9×10^{-7}
Collagen	345,000	6.9×10^{-8}
Catalase	250,000	4.1×10^{-7}
DNA	6,000,000	1.3×10^{-8}

Source: C. Tanford, *Physical Chemistry of Macromolecules.* (New York: John Wiley and Sons, Inc.: 1961).

7-7. Intuitive Meaning of Diffusion

Whenever we specify that at equilibrium there are no electrochemical potential differences in connected regions, we also imply that in any system which starts at different electrochemical potentials flows (of mass, heat, charge, etc.) will ensue such that eventually equilibrium will be reached and the electrochemical potential will be the same in all regions. Nature, then, abhors space differentiation: No preference is given to one region of space over another—provided, of course, that they look structurally alike. In the simple case in which we establish a concentration difference at constant T, P, for example, the solute will eventually distribute evenly in all regions it can reach. Thus, if a drop of ink is placed in a glass filled with water, ink will move through the vessel until the whole beaker acquires the color of the ink (the final color will, of course, be lighter).

What has happened? Two processes have taken place at once; from the point of view of the water molecules, it is the concentration of *water* that is different in different parts of the vessel so its electrochemical potential is larger in the more concentrated side than in the side that has the ink. Water, then, will move from the region in which its potential is large (pure-water part) to the region where its potential is low (the ink region), as discussed in Chapter 3 and illustrated in Fig. 7-5(a). The ink particles, on the other hand, "see" a large concentration where the ink is and a low concentration on the pure-water side. Their electrochemical potential will thus be higher in the ink region than in the pure-water region and hence will move toward the water side, as shown in Fig. 7-5(b). In the real world, the situation will not be this simple: Water molecules going in one direction will collide with ink particles going in the opposite direction—as will be discussed later in the context of the background provided by irreversible thermodynamics.

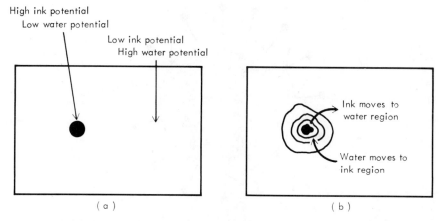

High ink potential
  Low water potential

        Low ink potential
        High water potential

Ink moves to
water region

Water moves to
ink region

(a)   (b)

Fig. 7-5. When a drop of ink is placed in water (*a*), ink moves to the water region and water moves to the ink region (*b*). This shows that chemical potentials of connected regions tend to be equated.

We should stress that diffusional forces are closely related to osmotic forces. If the solute (ink, for example) is prevented from moving to the water side—such as is the case when a selectively permeable membrane is introduced as a separation between the ink and the water regions—the solute will not move but water will "see" the difference in electrochemical potential, as discussed in Chapter 6, and will move from the pure-water side to the ink side, diluting the concentration of the ink in an attempt to equate both concentrations. If the volume of flow of water is restricted through the use of a fixed volume, an osmotic pressure will develop which will prevent the further flow of water.

7-8. Einstein's Molecular View of Diffusion

The physical connection between osmosis and diffusion was first established by Albert Einstein, who devoted some effort to explaining the phenomenon of Brownian motion. Botanist Robert Brown had observed in 1828 that pollen grains of different plants would move in an irregular manner when placed in water, and he attributed this phenomenon to the fact that the pollen grains had life. The same type of "dancing pattern" could also be observed with other large particles suspended in a solvent. More than ten different theories of Brownian motion appeared in the literature before Einstein showed in 1905 that it could be explained on the basis of the kinetic-molecular theory of heat and, furthermore, that diffusion and osmotic pressure were based on the same physical principle.

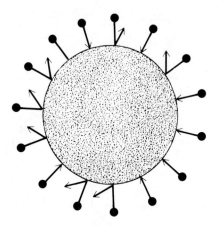

Fig. 7-6. If a large molecule or particle is suspended in solution, it will show a random motion because the smaller solvent molecules hit it at random from every direction.

In essence, Einstein's explanation was that the irregular motion of the molecules in the solvent—namely, water—leads to collision between the solvent and the solute (the suspended particles) which occurs in a completely random manner. At any given time there will be more molecules hitting in one direction than in another (as shown in Fig. 7-6). As a result, there is a net displacement of the suspended particles which is random in direction and follows a path mathematicians call a "random walk."

The random walk is typically exemplified by the walk of a drunkard who advances in every possible direction and takes steps of equal length. In the case of the suspended particles each of the steps has, on the average, a value Δ, which we must find for each solute-solvent system. Let us assume that each step takes an equal time τ. Consider what happens in that time in a simple system in which there are two compartments separated by a permeable partition of thickness Δ and maintained at different concentrations, c_1 and c_2 (Fig. 7-7).

If we pay attention merely to the motions to the left and to the right of the partition, we see that the molecules on the left of the partition, which are at a concentration c_1, will have equal probability of moving to the left as of moving to the right; we can assume that, on the average, half of these molecules will go through the partition while the other half will move away from it in time τ. The same argument can be given about the molecules on the right side. As a result, the net flow from left to right will be the difference between these two flows; that is,

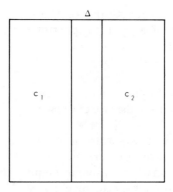

Fig. 7-7. A simple derivation of the molecular-diffusion properties across a thin membrane separating two compartments can be given by assuming that, on the average, the molecules in both compartments have an equal chance of moving to the left or the right in time, comparable to intermolecule collision times. The net flow to the right can then be given by $A/2 \, (c_1 - c_2) \, \Delta/\tau$ (see Eq. 7-14).

net flow of molecules to right = flow to right-flow to left.

But, the number of molecules in a given region will be the product of the concentration times the volume. Thus,

number of molecules moving to the right = $\frac{1}{2} c_1 \, A\Delta$

and   number of molecules moving to the left = $\frac{1}{2} c_2 \, A\Delta$.

The net flow to the right is, then,

$$(7\text{-}14) \qquad \frac{\text{number of molecules}}{\text{time}} = \frac{\frac{1}{2}(A\Delta)}{\tau} \, [c_1 - c_2],$$

in which Δ is the thickness of the membrane. Equating the two expressions, we obtain

$$\frac{1}{2} \frac{A\Delta}{\tau} \, [c_1 - c_2] = \frac{DA}{\Delta} \, [c_1 - c_2].$$

We can cancel out concentration differences and the area to obtain

$$\frac{1}{2} \frac{\Delta}{\tau} = \frac{D}{\Delta},$$

which yields the value of the average displacement

$$(7\text{-}15) \qquad\qquad \Delta = \sqrt{2D}\sqrt{\tau}.$$

That is, the average displacement observed depends on the time of observation, and is proportional to the diffusion constant. Notice that the displacement is not proportional to the time but to the *square root of the time.*

7-9. Evolutionary Pressures Imposed by Diffusional Processes: Diffusion and the Size of Organisms

Unicellular organisms must perform all the basic metabolic activities of digestion, respiration or fermentation, etc.; furthermore, most *unicellular* animals must also move to acquire their food, as they lack the appropriate photosynthetic machinery. If we correlate size and structure of unicellular organisms, it becomes clear that the larger the organism the more specialized the structures it acquires. In the case of the simple bacterium of the human intestine, *E. coli,* which measures about 2 microns (2/1,000 of a mm), there is little obvious specialization in the cell. Because of the small size of the cell, most materials can move quickly throughout the bacterium by diffusion. As we consider larger unicellular animals (Fig. 7-8), it is seen that some of the cells have acquired structures specialized in activities such as digestion or pumping of fluids for egestion (e.g., paramecium, euglena), because the law of diffusion cannot guarantee that the necessary metabolities will reach all parts of the cell in short times. Other cells have coped with this problem at the control level. Thus, the paramecium has two specialized nuclei, the macro- and the micronucleus, which control different activities.

How big are cells? Cells vary widely in size. If an average cell is approximated by a sphere—this is not quite valid—the radius of most cells ranges between 0,001 millimeter and 0,020 millimeter. A more convenient unit of length used in the measurement of cells is the micron, denoted by μ, which is a thousandth of a millimeter. The radius of most cells lies between 1 μ and 20 μ; an *E. coli* cell is 2 μ long and 1 μ wide. Other bacteria may be a tenth of this size, and a large ostrich egg is a hundred thousand times as large (20 cm). Figure 7-9 gives a comparison of cell sizes.

Evolutionary processes could have kept on increasing the size of *unicellular* animals indefinitely, but this would not have been an efficient way of tackling the problem of increasing overall size. In the case of an ideal spherical cell, the area available for diffusion is

$$A = 4\pi R^2,$$

while the volume that metabolizes is

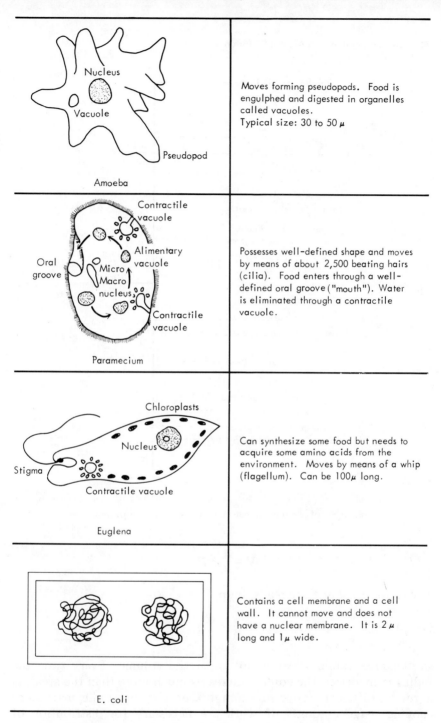

Amoeba	Moves forming pseudopods. Food is engulphed and digested in organelles called vacuoles. Typical size: 30 to 50 μ
Paramecium	Possesses well-defined shape and moves by means of about 2,500 beating hairs (cilia). Food enters through a well-defined oral groove ("mouth"). Water is eliminated through a contractile vacuole.
Euglena	Can synthesize some food but needs to acquire some amino acids from the environment. Moves by means of a whip (flagellum). Can be 100μ long.
E. coli	Contains a cell membrane and a cell wall. It cannot move and does not have a nuclear membrane. It is 2 μ long and 1 μ wide.

Fig. 7-8. Chart showing some typical unicellular organisms of different sizes and organization.

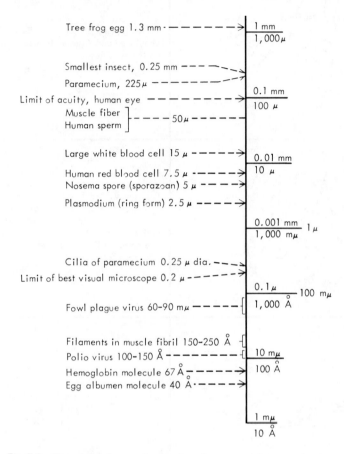

Fig. 7-9. Sizes of various animal cells, viruses, and large molecules.

$$V = \frac{4}{3} \pi R^3 .$$

The ratio of volume to area is, then,

(7-16)
$$\frac{V}{A} = \frac{R}{3} ,$$

so the area cannot "catch up" with the volume: Every time the radius is increased the volume increases much more than the area. As a result, diffusion processes cannot keep the metabolic machinery running; large amounts of work are necessary for distribution of metabolites, etc. The process of living becomes a very expensive commodity in terms of the amounts of energy needed to sustain it.

The evolutionary trend was, then:

- To reduce the size of cells as much as possible
- To divide the work of maintaining the metabolic machinery between many specialized cells

An additional way of reducing the amount of programming involved has been to give all the cells, even the highly specialized ones, the program to make the whole organism. Although this is an area of developmental biology which is poorly understood, it is clear that cells specialize not by acquiring new information or carrying very specific instructions at the start but precisely by the opposite process: They suppress part of the program and concentrate on a single task.

7-10. Why Are Cells So Large?

We have given a general indication as to why cells are so small. The question still remains as to why cells are *so large*; thus, even a typical *E. coli* bacterium contains billions of molecules grouped into from 3,000 to 6,000 different *kinds* of molecules. There are, of course, many molecules of each kind. Most of this variety, we should keep in mind, comes from attaching the same basic molecules in different combinations and varying their length. What is the most abundant molecule in any cell? This may come as a surprise to many; it is not lipid, protein, carbohydrate, or nucleic acid but the simple water molecule (to which life owes most of its properties). In *E. coli* there are about 40,500,000,000 molecules. Out of these, 40,000,000,000 are water molecules! Other chemical species present in *E. coli* are given in Table 7-2.

Why should there be so many molecules in a single cell? Physicist Erwin Shrödinger gave an answer that relates back to the previous considerations on the probabilistic nature of entropy as well as to the molecular reason for diffusion: Cells must be large enough to maintain a stable structure, otherwise they would not be able to work steadily and all their physical variables would be in a constant state of fluctuation.

7-11. The Problem of Fluctuations

Although the assumption that the same number of molecules move to the right as to the left is perfectly justifiable when a large number

TABLE 7-2. Principal Molecules in *E. coli*

COMPONENT	% TOTAL WEIGHT	NUMBER PER CELL	DIFFERENT KINDS
Water	70	40,000,000,000	1
Inorganic ions (Na$^+$, K$^+$, Mg^{++}, Ca^{++}, etc.)	1	250,000,000	20
Carbohydrates and related molecules (precursors)	3	200,000,000	200
Amino acids and precursors	0.4	30,000,000	100 (only 20 are finished amino acids)
Nucleotides and precursors	0.4	12,000,000	200 (only 200 are different nucleotides)
Lipids and precursors	2	25,000,000	50
Other small molecules	0.2	15,000,000	200
Proteins	15	1,000,000	2000 to 3000
Nucleic acids DNA	1	4	1
m RNA		1,000	1,000
tRNA		400,000	40

Source: J. D. Watson, *The Molecular Biology of the Gene* (New York: W. A. Benjamin, 1965).

of molecules are considered, it breaks down when there are only a few. The results of kinetic theory depend on averages and do not hold for a system with a few particles, because averages are not well defined in those instances. It is meaningless, for example, to talk about temperature when there are only a few molecules in a container: A thermometer placed inside the container will not, as a general rule, be able to hit enough particles to measure their average kinetic energy. Instead, the reading will fluctuate in time depending on which particles the thermometer is hit by (Fig. 7-10).

This behavior is typical of quantities that depend on probabilities or average values of small samples. If one asks for the probability of getting "heads" when a coin is flipped, the answer is "1/2" (or a 50/50 chance), but no one assures that there will be exactly five heads and five tails after flipping the coin only ten times (try it). On the other hand, if the same coin were to be flipped a million times, one could be almost sure that the number of heads would be about

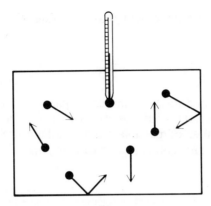

Fig. 7-10. A thermometer placed inside a container with a few particles will not be able to measure their energy because there will not be enough collisions to give a reliable measurement.

equal to the number of tails. A real experiment of coin flipping may look like Fig. 7-11. It can be seen that the probability oscillates around the central value "1/2" and that the fluctuations are larger the smaller the sample size.

For particles considered in the molecular diffusion phenomenon, we can define a ratio of the fluctuations of the number of particles in the given region to the total number of particles; this is a relative

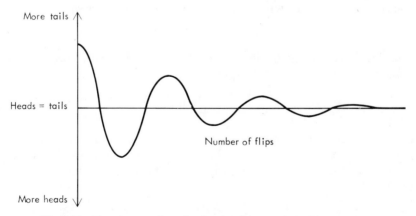

Fig. 7-11. The chance of getting heads when a coin is flipped is 50 percent, but the point at which the number of heads equals the number of tails occurs only after many flippings. For a small number of trials, there will be a fluctuation around the equal-heads and equal-tails value, whose amplitude decreases as the number of trials, N, increases and can be estimated from the expression $1/\sqrt{N}$. Similar fluctuations from average values are also seen in molecular events.

fluctuation, and it can be shown that the value of this ratio is (Prob. 7-6)

$$(7\text{-}17) \qquad \qquad \frac{1}{\sqrt{N}} \ .$$

Thus, the fluctuations in the number of particles in a given volume depend on the square root of the number of particles in the volume: The larger the number of particles, the smaller the fluctuation. This explains why an average cell contains so many molecules.

7-12. Gas Diffusion

Ideal gases, too, obey Fick's law of diffusion. In this case, however, the variable that dictates whether or not motion will take place is not the concentration of the gas but its *partial pressure*. In an ideal gas mixture each gas will contribute independently to the total pressure. If O_2 and CO_2, for example, are placed in a container of volume V, each gas will independently obey the ideal-gas law:

$$(7\text{-}18) \qquad \qquad VP_{O_2} = n_{O_2} RT \qquad \text{and}$$

$$(7\text{-}19) \qquad \qquad VP_{CO_2} = n_{CO_2} RT.$$

The *total* pressure, P, will be the sum of the partial pressures,

$$P = P_{O_2} + P_{CO_2}.$$

It is simple to visualize the relationship between partial pressure and concentration since, for example,

$$P_{O_2} = \frac{n_{O_2}}{V} RT$$

and, at a constant temperature, the partial pressure of oxygen will be proportional to the number of moles per unit volume—an equivalent "concentration." For this reason, the chemical potential of a gas i is defined as

$$(7\text{-}20) \qquad \qquad \mu_i = RT \ \ln P_i$$

or, for small differences

$$\Delta\mu_i = \frac{RT \ \Delta P_i}{P_i}$$

in which P_i is the partial pressure of the gas.

Since these expressions are exactly analogous to the chemical potentials considered in the derivation of Fick's law—except that P_i has been substituted for c—it follows that Fick's law for a gas is

$$J_i = \frac{A}{d} RT\omega \, \Delta P_i.$$

As before, the flow of the gas is proportional to its partial pressure difference, proportional to the cross-sectional area, and inversely proportional to the thickness of the partition, or membrane.

These considerations are useful in understanding gas exchange in the blood.

7-13. Respiration in Vertebrates as an Example of Gas Diffusion

The general scheme of fuel burning in vertebrates was given in Fig. 1-5. This method of gas exchange requires that oxygen and foods be continuously supplied to the tissues and that carbon dioxide and waste products be continuously removed. Some of these motions of matter are mediated by metabolic processes; these require pumping, and they are typical of food absorption in the intestine. The exchange of gases, on the other hand, is purely diffusional and proceeds spontaneously from regions of high gas pressure to regions of low gas pressure.

Since oxygenated blood coming out of the lungs is carried by arteries and deoxygenated blood is returned through the veins (Fig. 7-12), it follows that in order for O_2 to flow spontaneously to the

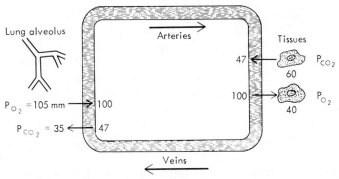

Fig. 7-12. Simplified diagram showing how oxygen and carbon dioxide flow passively between tissues, blood, and lungs. P_{O_2} and P_{CO_2} are the partial pressures of oxygen and carbon dioxide, respectively.

tissues the partial pressure of oxygen in the lungs must be larger than the partial pressure of O_2 in the arterial capillaries. The latter should, in turn, be larger than the oxygen partial pressure in the tissues. This is indeed the case:

	LUNGS (ALVEOLI)	ARTERIES	TISSUES
P_{O_2}:	105 mm	100 mm	40 mm

in which pressures are given in terms of the equivalent height of mercury (in mm) needed to obtain the same pressure. In the case of CO_2, the pressure of CO_2 in the tissues should be *larger* than the pressure of CO_2 in the vein capillaries, which, in turn, should be larger than the CO_2 pressure in the lungs. This is, again, what happens,

	TISSUES	VEINS	LUNGS
P_{CO_2}:	60 mm	47 mm	35 mm

It is clear, then, that the motion of O_2 is

$$lungs \rightarrow blood \rightarrow tissues,$$

while the motion of CO_2 is

$$tissues \rightarrow blood \rightarrow lungs$$

and that the basic motion of gases during respiratory exchanges can be explained in terms of simple diffusional processes—in terms of spontaneous processes from high to low chemical potentials. We should point out, however, that there is more to the process of respiration than these simple considerations of partial pressures.

First, respiratory gases like O_2 and CO_2 do not dissolve in the blood directly (if this were the case, 100 ml of blood would carry only 0.3 ml of O_2) but associate with hemoglobin, the molecule that gives blood its red color. Hemoglobin is carried by specialized wandering cells, the red blood cells; an average red cell carries 30 milligrams of Hb. O_2 and CO_2 can pass freely through the cellular membrane. In the lungs the tubes that carry air ramify many times and make contact with very small blood vessels, the pulmonary capillaries, which admit only one red cell at a time. The total area for exchange of oxygen and carbon dioxide is large—on the order of 100 square meters. According to Fick's law, a large area increases the rates of diffusion, so a given red cell need not be stationed at the lungs for long to effect the exchange. Since the capillaries are so fine, each red blood cell is in close contact with the walls of the capillaries—hence close to the tissues that require O_2 and CO_2. The distance d, which appears in Fick's law, is, therefore, small, and the resultant flow is large. In general, cellular regions through which fast transport of matter is required have large areas. Sometimes this increase is achieved through specialized infoldings, the *microvilli* (Fig. 7-13).

7-14. Some Unrelated Transport Processes

Diffusion is not the only possible way for a molecule to move through a partition such as a biological membrane. One of the interesting aspects of the diffusion concept is, surprisingly, that it

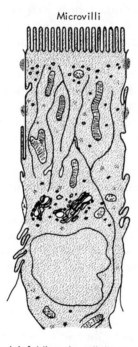

Microvilli

Fig. 7-13. Specialized infoldings in cellular membranes (such as the intestinal microvilli shown) increase the area available for passive or active transport of solute, and are found in regions where large flows take place. The large number of mitochondria in this cell indicates that active transport is taking place. (C. P. Swanson, *The Cell.* Englewood Cliffs, N.J.: Prentice-Hall, Inc., 1960.)

can tell what the system *is not* doing; in particular, it can give an indication as to whether diffusion or a different process is taking place. In some experimental situations this can be as informative as finding out what the system is doing.

For example, if we observe matter moving from a dilute to a concentrated compartment at the same temperature and pressure, we can exclude simple diffusion as an explanation for the mechanism involved. Since motion against a concentration gradient at constant temperature and pressure implies, when there is only one type of species moving, that metabolic energy is being spent, one would need to postulate the existence of some kind of metabolic "pump." If there are several substances diffusing, however, there may be some interaction among the diffusional flows of the various substances, so we cannot conclude immediately that a pumping process is present. We can group all processes in which metabolic energy is being spent as "active transport" and those in which the driving energy is concentrated in electrochemical potential gradients (differences in concentration, pressure, etc.) as "passive processes." Fick diffusion is the most typical example of passive diffusion. Other examples are given in Fig. 7-14. To a first approximation, we can say that if a substance moves *against* its own electrochemical potential gradient, the process is active. There are exceptions which are considered by nonequilibrium thermodynamics, but for all practical purposes this approximation suffices. Hans Ussing has given a practical test of active transport based on measuring ratios of unidirectional flows. This method is useful when a single solute is placed in two compartments separated by a membrane and the solute in each compartment is labeled with a different radioactive atom, so that as the solute molecules move in both directions, the flows from left to right and right to left can be individually followed. By assuming that the unidirectional flows do not interact, Ussing was able to show that their ratio during passive diffusion of an uncharged solute is related to the ratio of the concentrations in the compartments by

$$\frac{\overrightarrow{J}}{\overleftarrow{J}} = \frac{c_1}{c_2}$$

in which c_1 and c_2 are the left and right concentrations, respectively, and the arrows indicate the direction of flow. If the ratio of unidirectional tracer flows is plotted against the ratio of concentrations, a straight line is obtained which passes through the point (1,1)—when the concentrations are equal in both compartments, the two flows are equal. This is also the case in other passive processes studied by Ussing. During *facilitated transport* the

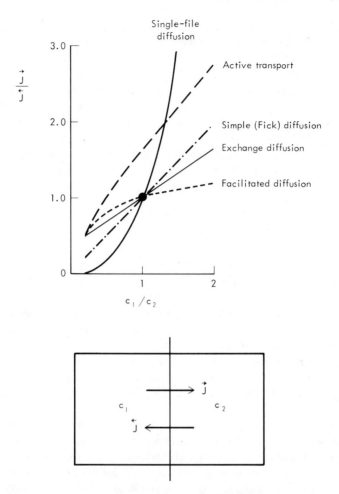

Fig. 7-14. Method devised by physiologist Hans Ussing to distinguish diffusional flows from active transport when *only one substance* is present, in various concentration ratios across a partition. The ratio of unidirectional flows is always one for all passive processes in which the concentration of the solute is the same in both compartments. The unidirectional flows may be measured by utilizing radioactive tracers, and are indicated as $\overrightarrow{J}$ or $\overleftarrow{J}$, depending on whether the unidirectional flow goes from left to right or right to left. (P. F. Curran and S. G. Schultz, *Handbook of Physiology*, sec. 6, vol. III, p. 1217.)

solutes move across the membrane faster than predicted by Fick's law but they never move against their electrochemical potential difference. In the case of *exchange diffusion*, a molecule of the solute on the right is exchanged with a molecule of the solute on the left so that there is no net flow.

Unlike facilitated transport, simple (Fick) diffusion or exchange diffusion, in the case of active transport the plot of flow ratios vs. concentration ratios does not go through the point (1,1) and there is unidirectional net flow when the concentrations in both compartments are equal. This pumping process requires metabolic energy. In regions in which active transport takes place, large numbers of mitochondria and ATPases—the enzymes that catalyze the reaction ATP $\longrightarrow$ ADP—are found.

We consider the mammalian kidney as an example in which passive diffusion, diffusion controlled by a hormone, and active transport take place at the same time.

7-15. The Nephron

The mammalian kidney is composed of thousands of filtering structures called *nephrons* [Fig. 7-15(a)]. Each nephron is very convoluted, as shown in the figure, and can be thought of as consisting of a coarse filter, the *glomerulus*, and three parallel tubes [shown from left to right in Fig. 7-15(b)]: the proximal tubule, the distal tubule, and the collecting duct. The glomerulus allows small and medium molecular weight substances to go into the tubules, while it retains large proteins and cells. These larger elements return to the main circulation through a blood capillary system present in the glomerulus.

The basic job of the nephron consists in getting rid of noxious substances such as nitrogen compounds (ammonia, urea, and uric acid) and recovering useful substances (glucose, amino acids, and especially water). A steady state is maintained in the nephron [Fig. 7-15(c)]: as fluid goes down the proximal tubule, the concentration of Na^+ increases (because Na^+ is pumped out of the distal tubule into the proximal tubule). The maximum concentration occurs at the convolution which joins the proximal with the distal tubules, the Loop of Henle. As the fluid goes up the distal tubule, the concentration of Na^+ decreases (because Na^+ is being pumped out of the distal tubule). Water would normally flow out with the salt as an osmotic gradient is established by the pump; the distal portion of the tubule, however, is impermeable to water. The minimum concentration of Na^+ occurs at the end of the distal tubule and beginning of the collecting duct. The collecting duct is permeable to water, and water moves out of the duct with a flow proportional to the concentration difference of Na^+ between the distal tubule and

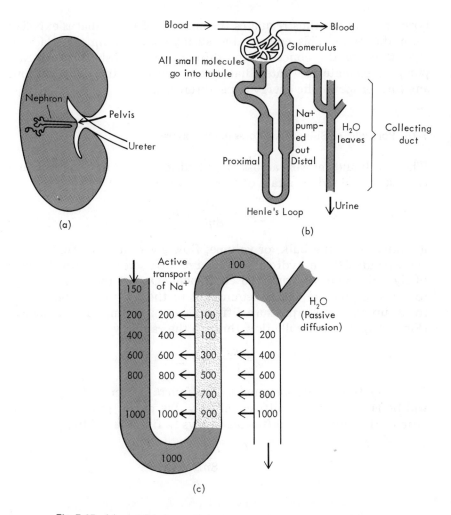

Fig. 7-15. (a) and (b) General architecture of the mammalian kidney showing the nephron, the basic functional unit. (c) The kidney could effect a concentration of salts by pumping H_2O out. (d) Instead, it pumps out Na^+ and then removes H_2O by passive diffusion. The antidiuretic hormone (vasopressin) regulates the amount of water retained by the kidney. (Based on K. Schmidt-Nielsen, *Animal Physiology*. Englewood Cliffs, N.J.: Prentice-Hall, Inc., 1960.)

the collecting duct. Water leaves from the collecting duct throughout so that as the fluid moves down the collecting duct the urine becomes more and more concentrated in Na^+. The permeability of the collecting duct to water is controlled by a hormone (the diuretic hormone), which thus regulates the amount of water that is expelled from the body. This complicated steady-state system takes advantage, then, of the concentration differences established by the Na^+ pump to regulate the water that is removed from the body without any further metabolic energy expenditures.

7-16. Poiseulle, or Bulk, Flow: Pores in Membranes

When a liquid moves across a cylindrical pipe of length l and cross-sectional radius, a rate of flow is given by

(7-21)
$$J_v = \frac{\pi a^4}{8\eta l}\, \Delta P,$$

in which J_v is the bulk, or volume, flow given in cubic centimeters per second, ΔP the hydrostatic pressure difference between the ends of the tube, and η a frictional coefficient. One of the possible ways to create a pressure exactly equivalent to the hydrostatic pressure is to set up an osmotic pressure difference using an impermeant species (Fig. 7-16). The Poiseulle flow through the capillary is then

$$J_v = \frac{\pi a^4}{8\eta l}\, RT\, \Delta c.$$

Since this flow is proportional to the *fourth* power of the radius, it will be larger than the diffusional flow. We can in principle find the ratio of the bulk to the diffusional flows to find the radius a,

$$\frac{J_v}{J_D} = \frac{RTa^2}{8\eta D}\, .$$

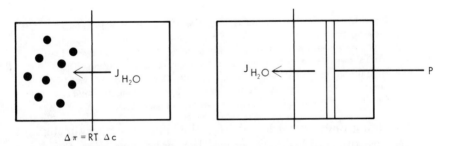

$\Delta \pi = RT\, \Delta c$

Fig. 7-16. A water flow can be established by either a hydrostatic pressure or an equivalent osmotic gradient.

7-17. Applications to Biological Membranes

Some epithelial tissues such as the toad bladder and the frog skin appear to be perforated by water-filled channels which constitute the passage route for water and water-soluble substances. It is then possible to compare the diffusional and the bulk flows to gain some idea of the approximate size of the "channels." Although the figures obtained depend on several assumptions, including the fact that channels should be circular cylinders, the calculation does give some general information as to porosity of the tissue.

Physiologist Alexander Leaf used this information to calculate the size of these "pores" in the toad bladder and obtained a value of 40 angstroms. But this value cannot be taken too seriously because the electron microscope does not show any visible pores this size.

There are some indications that cellular membranes also have such openings. Most of the cellular membrane is composed of lipids through which water diffuses slowly; as a result, diffusion of lipid soluble substances increases with increasing lipid solubility (Fig. 7-17). Water motion across cellular membranes under the influence of a hydrostatic pressure is much larger than simple diffusional flow (Fig. 7-18), suggesting a Poiseulle type of flow and the presence of water-filled pores. Arthur K. Solomon and colleagues of Harvard Medical School have measured the equivalent pore size of red blood cells using various flow methods. They estimate that the *equivalent* radius of an average pore is of the order of 4 angstroms, in a red cell.

7-18. Interaction Between Osmotic and Hydrostatic Gradients

Poiseulle's law for bulk motion of fluid under a hydrostatic pressure (Eq. 7-21) can also be written as

$$(7\text{-}22) \qquad\qquad J = L\,\Delta P,$$

in which L, the *hydraulic conductivity* is given by

$$L = \frac{\pi a^4}{8\eta l}$$

and ΔP is the pressure difference between the ends of the tube through which the bulk flow takes place. If, *in addition* to the hydrostatic pressure ΔP there is a concentration difference of a species which cannot go through the "tube," or "pore" (Fig. 7-19),

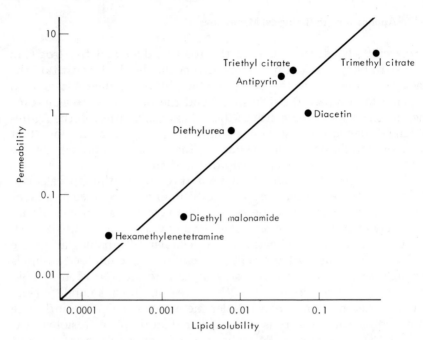

Fig. 7-17. Permeability (in cm/hr) of the plant cell *Chara cerato-phylla* to molecules of similar size, as a function of their lipid solubility. The abscissa is the olive oil in water partition coefficient. (Adapted from Collander, *Phys. Plant., 2*, 300, 1949.) The graph shows that the molecules of the same size go more easily through a cell membrane the more lipid soluble they are.

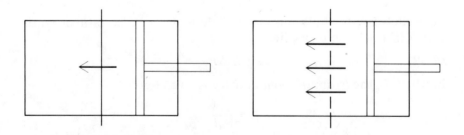

Fig. 7-18. The flow of water across a membrane which is not very permeable to water will be increased if there are water-filled pores in the membrane. In that case a pressure difference will give rise to a Poiseulle flow.

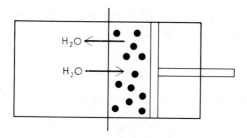

Fig. 7-19. If both a hydrostatic pressure and an osmotic pressure are present, their individual contributions add up.

the *effective* pressure difference will be $\Delta P - RT \Delta c$ and the total expression for the flow will be

$$(7\text{-}23) \qquad\qquad J = L\,(\Delta P - RT\,\Delta c).$$

What happens if the tube is "leaky" and some or all of the molecules which contribute to the osmotic pressure difference go through? In this case the reduced osmotic pressure is accounted for by introducing the reflection coefficient σ (Sec. 6-17). The resultant volume flow is

$$(7\text{-}24) \qquad\qquad J = L\,(\Delta P - \sigma RT\,\Delta c),$$

It is usually convenient to consider the volume flow per unit area, J/A. We shall denote this flow by Jv,

$$(7\text{-}25) \qquad\qquad J_v = L_p\,(\Delta P - \sigma\,RT\,\Delta c),$$

in which L_p, the hydraulic conductivity coefficient, is the ratio of the volume (bulk) flow to the pressure difference when the concentration difference is zero.

7-19. In Dilute Solutions J_v is the Velocity of the Solvent

In all practical cases of biological interest, volume changes are caused by the entry or exit of water. The volume flow J_v, then, is essentially the volume of water that goes through a unit area per unit time. Strictly speaking, this is true only for dilute solutions, as shown in the Appendix. For the time being, we shall write $J_v = J_w / A$, in which J_w / A is the volume of water per unit area.

The flow of water per unit area is

$$(7\text{-}26) \qquad\qquad J_v = \frac{J_w}{A} = \frac{v_w}{At},$$

in which Vw is the volume of water and t the time it takes for this volume to go across the area. The water volume can also be written as

(7-27) $$V_w = A \times d,$$

in which d is the thickness of the region and A the cross-sectional area. Equation 7-26 then becomes

(7-28) $$J_v = \frac{Ad}{At} = \frac{d}{t}.$$

But the distance traveled by water per unit time is simply v_w, the water velocity. Thus, we obtain the equality

$$J_v = v_w.$$

We then have, from Eq. 7.25,

(7-29) $$v_w = L_p \, \Delta P - L_p \sigma RT \, \Delta c.$$

7-20. Hydrostatic Pressures "Push" Solute Molecules

We have so far looked only at the effect that concentration differences have on volume flows. It should be clear that when hydrostatic pressures are applied across a partition *solute* molecules will also be dragged along if they can go through the partition at all. What is the relationship between solute and volume flows?

When all solute molecules can go through the membrane ($\sigma = 0$), the velocity of the solute molecules v_s equals the velocity of the water molecules; that is,

(7-30) $$v_s = v_w,$$

so the solute flow per unit area is given by

(7-31) $$J_s = \overline{c_s} v_s = \overline{c_s} v_w,$$

in which $\overline{c_s}$ is the (average) concentration of solute molecules in the partition or membrane. Since $v_w = J_v$, we obtain

(7-32) $$J_s = \overline{c_s} J_v$$

for the special case of zero reflection coefficient. In the other extreme case, $\sigma = 1$, all solute molecules are "reflected" and

(7-33) $$J_s = 0.$$

In general, the solute flow will be proportional to the fraction of solute molecules *transmitted*. This is given by fraction "transmitted" = 1 – fraction "reflected"

$$= 1 - \sigma.$$

The solute flow under hydrostatic pressure is, then,

(7-34) $$J_s = (1 - \sigma)\bar{c}_s J_v.$$

Since J_s/c_s is v_s, the solute velocity, we can also write Eq. 7-34 as

(7-35) $$v_s = (1 - \sigma)J_v.$$

The meaning of this expression is that it is possible to obtain a diffusional flow of solute (even when there are no concentration gradients) as a result of the application of a hydrostatic pressure. The solute is dragged along with the solvent, and this is one of the possible cases in which *passive* motion, which is independent of metabolic "pumps," can be confused with "active" transport (Fig. 7-20). Thus, Harold Andersen and Hans Ussing showed that two solutes, acetamide and thiourea, could be dragged across the toad skin between two compartments at the same concentration by an induced water flow (Fig. 7-21).

7-21. Diffusional Flow in the Presence of Hydrostatic and Concentration Gradients

Equation 7-35,

$$v_s = (1 - \sigma)J_v,$$

leads directly to a slightly different expression,

(7-36) $$v_s - J_v = -\sigma J_v$$

or, recalling $J_v \cong v_w$,

(7-37) $$v_s - v_w = -\sigma J_v$$

and, since $J_v = L_p \Delta P$ in the presence of a hydrostatic pressure difference, we obtain

(7-38) $$v_s - v_w = -\sigma L_p \Delta P.$$

This velocity difference gives the velocity of the solute relative to water, and it is a form of diffusional flow. While our original derivation of Fick's law considered only the motion of solute, water too will move in the opposite direction (Fig. 7-22). We can denote the velocity difference by J_D, the diffusional flow per unit area. If a concentration difference is also present, it contributes to the

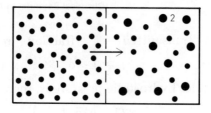

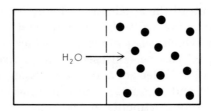

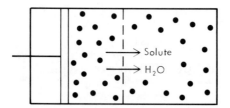

Fig. 7-20. According to equilibrium considerations, if a solute moves between two compartments at the same temperature, pressure, and concentration of a given substance, work must be done. In some cases, however, an incomplete description may lead to the incorrect conclusion that "active transport" is taking place. Thus, if we were aware only of the small molecules in the drawing, it would look as though the system were somehow pumping these molecules from one compartment to another. The second drawing shows that the larger molecules, which cannot go through the membrane, are actually a source of an osmotic force which causes water to enter the compartment. This motion couples to the diffusional flow of the solute so that the solute is pulled along with the water molecules.

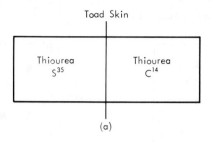

(a)

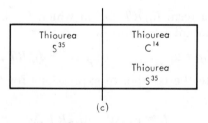

(b)

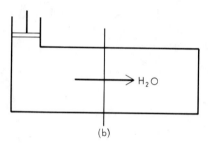

(c)

Fig. 7-21. (*a*) Two different isotopes of thiourea—thiourea molecules having two different radioactive atoms, S and C—are placed in the two chambers at the same concentration. (*b*) A pressure difference is applied (actually, the pressure is osmotic rather than hydrostatic). (*c*) Thiourea S^{35} is found to appear in the compartment at the right. The amount of S^{35} detected is directly related to the amount of water that has passed through the skin. (B. N. Anderson and Hans Ussing, Acta Phys. Scandinavica, *39*:228.)

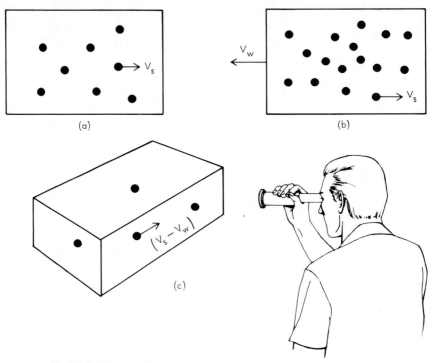

Fig. 7-22. (a) Relative to water, a solute molecule moves with velocity v_s. (b) Water molecules move in the opposite direction with velocity v_w. (c) Then to an observer in the laboratory, the molecule seems to move slower, with velocity $v_s - v_w$.

diffusional flow a term $L_D RT \, \Delta c$, in which L_D is a proportionality constant. The total diffusional flow J_D is then

$$(7\text{-}39) \qquad J_D = v_s - v_w = -\, \sigma L_p \, \Delta P + J_D \, RT \, \Delta c.$$

We can now write the composite expressions for the diffusional and bulk flows as

$$(7\text{-}40) \qquad J_v = L_p \Delta P - \sigma L_p RT \, \Delta c$$

$$J_D = -\, \sigma L_p \Delta P + L_D \, RT \, \Delta c.$$

While it is not surprising to find that the term $-\sigma L_p$ appears twice—once multiplying the term $RT \; \Delta c$, the second time multiplying ΔP—this result indicates (see problems at the end of the chapter) that the same change is obtained in the volume flow when a concentration difference is applied as the effect on the diffusional flow when a hydrostatic pressure is applied (Fig. 7-23).

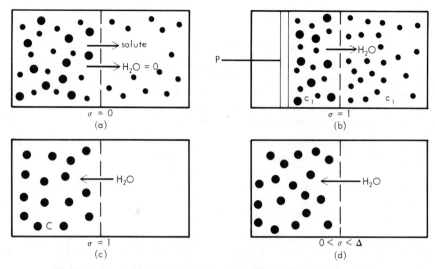

Fig. 7-23. The reflection coefficient σ provides an idea of the magnitude of the coupling between the motion of volume under the influence of a hydrostatic pressure and the diffusional flow caused by a difference in concentration of solute. When $\sigma = 0$, there is no coupling between the two processes, so a difference in concentration does not elicit the volume of water in the steady state, and a difference in hydrostatic pressure does not cause a motion of solute relative to the water molecules—a diffusional flow. Solute and water molecules move, in that case, at the same velocity. When $\sigma = 1$, we have an extreme example of undirectional coupling in which a difference in concentration of the solute molecules causes only a motion of water, but no solute flow. For other values of σ, there is always some coupling. Rigorously speaking, the water flow can be equated with the volume flow and the diffusional flow with the solute flow when the solute concentrations are small.

This result can be obtained from the more general considerations of nonequilibrium thermodynamics, as indicated in Chapter 11 and the Appendix.

Problems

7-1. Consider a uni-univalent electrolyte—a salt such as Na Cl which gives ions with one positive charge (Na^+) and ions with one negative charge (Cl^-) in solution. If a membrane permeable to both ions separates two compartments at which the salt is at different concentrations, a flow will occur from the concentrated to the dilute compartment.

(a) Write the steady-state expression for the flow of positive ions (cations), J_+, in terms of the mobility of the cations, W_+, the difference in electrochemical potential, $\Delta \overset{\curvearrowright}{\mu}_+$, the thickness of the membrane, d, and the concentration of salt in the membrane, c.

(b) Repeat for the steady anion flow, J_-.

(c) Give the explicit expressions for (a) and (b) in terms of concentrations and potential differences in both compartments.

(d) Recalling that the charge is conserved, give the potential difference between the two compartments.

(e) What is the diffusion constant for this flow?

7-2. Given the system above,

(a) What is the potential of the concentrated relative to the dilute compartment if the positive ions (cation) move faster than the anion?

(b) What is the potential if the mobility of the anion is larger than that of the cation?

(c) What is the potential difference if $\omega_+ = \omega_-$?

7-3. Cellular electrical potentials are measured by placing a very fine-tip pipette filled with concentrated K Cl inside the cell and measuring the electrical potential in the liquid. Knowing that the mobilities of K^+ and Cl^- are equal, explain why the pipette is filled with concentrated K Cl.

7-4. What is the meaning of the negative sign in Fick's equation?

7-5. Utilize the approach of Sec. 9-1 to show that the force exerted on a charged molecule is

$$\frac{F}{n} = \Delta \overset{\curvearrowright}{\mu} / d,$$

in which $\Delta \overset{\curvearrowright}{\mu}$ is the difference in electrochemical potential across two regions separated by a distance d, and n is the number of moles of the substance.

7-6. *Fluctuations.* We wish to show that the fluctuations discussed in Sec. 7-13 decrease as the inverse square root of the number of events by analogy with the fluctuations in coin flipping. Follow the following steps:

(a) Express the total number of flips (or steps, or events), N, in terms of the total number of heads, H, and the total number of tails, T.

(b) Define the displacement from the outcome of equal heads and tails after N trials as

$$D_N = H - T;$$

give D_N in terms of D_{N-1}. (There are two answers.)

(c) We are only interested in the absolute value of the displacement from the "half heads-half tails" position, regardless of whether there are more heads or tails. Square both answers in (b) to obtain the average square value of the displacement, D_N^2.

(d) Use the formula in (c) to find the squared displacement after 1, 2, 3 coin flippings. What is the displacement after N flippings?

(e) Take the square root of the result in (d) to find the average D_N. Relate this answer to (a) and (b) to give the final answer in terms of N.

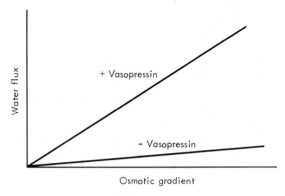

Fig. 7-24.

7-7. Figure 7-24 shows the general behavior of the water flow induced by an osmotic gradient across a toad bladder both in the presence and absence of vasopresin (a hormone). Is the increase in flow that obtains when vasopresin is added active or passive?

8

INFORMATION THEORY, CODES, AND MESSAGES

Biological expression takes the form of "messages" which contain information about structure and function. The shapes of living matter (structure) and the chemical reactions of life (function) are completely determined by the structural proteins, which take part in the static structural and dynamic architecture of the cell, and the enzymes, which catalyze the reactions involving small organic molecules. Once the organism knows how to make the proteins it needs and is able to keep the knowledge for future generations, it can metabolize and reproduce: it lives. The instructions on how individual proteins must be built are contained in genetic messages, the genes.

8-1. Information Theory

The general problem of transmitting and interpreting (decoding) messages is considered by information theory, a close relative of thermodynamics, which, a little by design and a little by chance, uses the statistical concept of entropy as a starting point.

In the general communication problem considered by Claude Shannon, the inventor of information theory, the following basic elements are introduced (see Fig. 8-1):

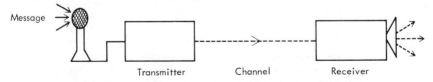

Fig. 8-1. Basic elements of a simple communication system: The message (voice) is encoded by the transmitter, sent through the channel, and decoded by the receiver.

- A *message*
- A *transmitter*: the thing that is sending the message
- A *receiver*: the instrument that reads and decodes the message
- A *channel*: medium through which the message is transmitted
- A *code*: or set of symbols used to write the message
- *Noise*: an undesirable signal that interferes with the whole process and cannot be eliminated

8-2. The Primitive Telegraph

A simple example is provided by the telegraph (Fig. 8-2). There is a code given by a sequence of lines, dots, and periods of silence; a transmitter, which serves to send the message in the form of an electromagnetic signal; a channel—the air; a receiver, which includes the operator who decodes the message. Noise is distributed throughout: There may be electrical discharges interfering with the real signal, errors caused by the operator, etc. In devising his dot-and-dash code Morse followed the principle of using the shortest symbols—the fastest to transmit—for the most common letters. This method is still used in more sophisticated codes.

8-3. The Genetic Message, a Biological Example

In the translation of genetic information the problem is complicated by the presence of several steps which can be viewed as a series of

Letter	Code	Count		Letter	Code	Count
E	·	12000		M	— —	3000
T	—	9000		F	· · — ·	2500
A	· —	8000		W	· — —	2000
I	· ·	8000		Y	· · — · ·	2000
N	— ·	8000		G	— — ·	1700
O	· ·	8000		P	· · · · ·	1700
S	· · ·	8000		B	— · · · ·	1600
H	· · · ·	6400		V	· · · — —	1200
R	· · · ·	6200		K	— · —	800
D	— · · ·	4400		Q	· · — ·	500
L	————	4000		J	· · · · ·	400
U	· · —	3400		X	· — · ·	400
C	· · ·	3000		Z	· · · —	200

Fig. 8-2. Morse's original code assigned the shortest string of symbols to the most common letters so as to minimize the transmission time of a typical message. Numbers represent times a letter appeared in a short English text.

receivers and transmitters; or, alternatively, the intermediate receivers and transmitters can be included in the channel.

Protein structure is carried in coded form in the DNA molecule; we recall that a DNA molecule is composed of a sugar-phosphate backbone with the bases adenine (A), thymidine (T), guanine (G), and cytosine (C). The linear array of these bases specifies the order and kinds of amino acids in the enzyme, or structural protein, each group of three bases coding for a single amino acid in the protein molecule. We can refer to the sequence of nucleotides coding for proteins as a message or, more rigorously speaking, a coded message. Instead of making the protein directly, the genetic message merely specifies which amino acid is to be assembled to which other amino acid, and the message is later transcribed into an RNA molecule and read like a recording tape by the factory assembly line, the ribosome, where the linear array of amino acids is put together to form the protein (Fig. 8-3).

As we have seen in Chapter 5, once the linear sequence of amino acids is specified, the three-dimensional shape of the protein is automatically determined, because the molecule folds into the shape of minimum energy, in which the total energy of repulsion and attraction between adjacent atoms is the minimum allowed to any possible configuration. Thus, the original coded message is interpreted to mean a given shape or function by expressing it in the form of a protein.

It is enlightening to point out some of the distinctive features that distinguish the genetic mechanism from a simple man-made system like the Morse telegraph.

First, the genetic "code" is not read off, or decoded, in the same way it was sent, but is *translated* in a peculiar manner: The language effectively changes during the process of transmission, even though the transformed words carry the same messages. During translation, the strand of DNA containing the genetic information is copied in the form of a complementary RNA molecule similar to the complementary DNA strand that holds the DNA structure together, with the added difference that the base uracyl (U) replaces thymidine (T) in RNA. This copy is very much like a negative of a photographic film: It contains the information but in reverse. Since this RNA molecule, the messenger RNA, or m-RNA, has more to do with the process of transmitting information than the original DNA, it is customary to consider the code in m-RNA as being the original message. There are four nucleotides in m-RNA which can form 64 different m-RNA code words, or *codons*, when taken in groups of three at a time (Table 8-1).

TABLE 8-1. The Amino Acid Code

UUU Phenylalanine	UCU Serine	UAU Tyrosine	UGU Cysteine
UUC Phenylalanine	UCC Serine	UAC Tyrosine	UGC Cysteine
UUA Leucine	UCA Serine	UAA Chain Termn.	UGA Chain Termn.
UUG Leucine	UCG Serine	UAG Chain Termn.	UGG Tryptophan
CUU Leucine	CCU Proline	CAU Histidine	CGU Arginine
CUC Leucine	CCC Proline	CAC Histidine	CGC Arginine
CUA Leucine	CCA Proline	CAA Glutamine	CGA Arginine
CUG Leucine	CCG Proline	CAG Glutamine	CGG Arginine
AUU Isoleucine	ACU Threonine	AAU Asparagine	AGU Serine
AUC Isoleucine	ACC Threonine	AAC Asparagine	AGC Serine
AUA Isoleucine	ACA Threonine	AAA Lysine	AGA Arginine
AUG Methionine	ACG Threonine	AAG Lysine	AGG Arginine
GUU Valine	GCU Alanine	GAU Aspartic acid	GGU Glycine
GUC Valine	GCC Alanine	GAC Aspartic acid	GGC Glycine
GUA Valine	GCA Alanine	GAA Glutamic acid	GGA Glycine
GUG Valine	GCG Alanine	GAG Glutamic acid	GGG Glycine

The final specification of a given amino acid is not achieved directly by the m-RNA molecule but through a small "adapting" molecule, transfer RNA (t-RNA), which carries a specific amino acid covalently bound which recognizes the codon in m-RNA through an "anticodon." This process of successive translations can be compared with the situation in which an Englishman communicates with a Russian by having two translators in between, one who understands English and speaks Chinese and another who understands Chinese and speaks Russian. Notice that reverse translation is impossible in this scheme. The process, which can be schematically specified by the irreversible chain of events

$$DNA \longrightarrow m\text{-}RNA \longrightarrow protein$$

is called the *central dogma* of molecular biology. Although some recent evidence seems to indicate that the process may be reversible in special cases, the central dogma stands as one of the most powerful unifying principles in biology.*

A *second* difference between the amino acid code and the simple Morse code is that the genetic code is *degenerate*: The same amino acid may be specified by different codons, as Table 8-1 shows. It is then surprising that in spite of this degeneracy the code is *universal*: Although the original codons were determined in the bacterium *E. coli*, all the organisms so far tested are able to interpret the messages the same way.

*Some cancer viruses contain an enzyme, the reverse transcriptase, which achieves DNA polymerization using DNA as a template.

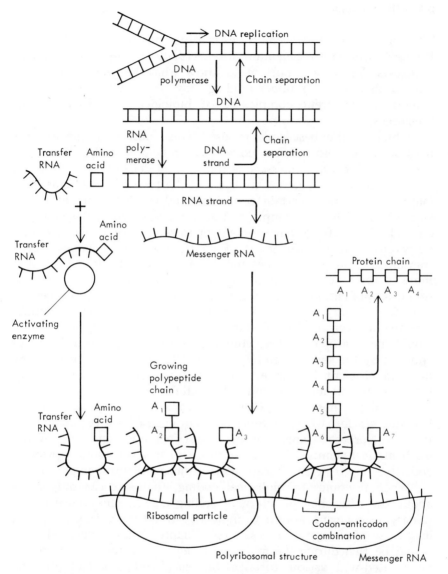

Fig. 8-3. Main events in the process of transmission of genetic information: The DNA message is translated into an RNA copy and read by the ribosome where the protein specified by the gene is assembled. (A. Rich, in *The Neurosciences*, ed. by G. C. Quarton et al. New York: Rockefeller University Press, 1967.)

8-4. Efficient Coding

The elaboration of efficient codes is a job that belongs to the James Bonds—or rather to their more intellectually inclined chiefs—not to biologists. Yet it is important to know what the general principles are, so that a better understanding may be gained of the reasons behind the apparent complexity of biological systems and their "messages."

Which is the best code to use? This question is meaningless because there is no absolute code that can be used under all possible conditions; the way a message is transmitted depends largely on the type of transmitter-channel we use and the type of noise. One code may be ideal under certain conditions and terrible under others. It would be foolish, for example, to transmit a message by flashing light on and off in a foggy environment or to send a sound message in a noisy environment. We could of course use large energies in the transmission process thus beating the noise by brute force, but it is more clever to use the flashlight in the noisy environment and the sound amplification in the fog (Fig. 8-4). Similarly, the process of biological energy conversion could consist of burning foodstuffs directly in oxygen and utilizing energy in the form of heat, but it would be inefficient. The complex series of intermediate steps adopted by the aerobic cell to generate ATP and use chemical energy does an efficient job under the restrictive conditions of constant temperature and acidity found in the cellular environment.

Although these are extreme examples, there is no way of knowing *a priori* what *the most efficient* code for a given channel is. It took Norbert Wiener, inventor of cybernetics and one of the mathematical geniuses of our century, to show what some optimum codes for very restricted forms of electromagnetic noise might be.

When it comes to biological coding, then, we can only *guess* about the efficiency of alternate coding mechanisms. Since evolutionary mechanisms appear to indicate a trend toward optimization in the use of energetic processes (see Chapter 10), an educated guess would be that during the millions of years of evolution, Nature experimented with various forms of biological coding and adopted the present mechanisms of energy conversion and genetic coding as the most efficient possibilities available consistent with the existence of life.

8-5. Optimum Man-made Codes; Binary System

There are some general principles of coding worth taking into account. These are all based on the fact that a transmitted message

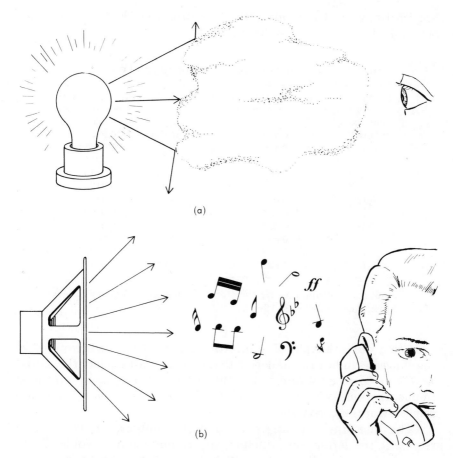

Fig. 8-4. The message must be sent in such a way that it will encounter the least possible interference. The figures show two *inefficient* transmissions: utilizing light in a hazy environment (*a*) and sound in a noisy environment (*b*).

usually contains more than one signal, or word, and that the signals appear with different frequency. Furthermore, each word takes, in general, different lengths of time to transmit. The total average time required to transmit a message will be given by

$$t_{\text{expected}} = \Sigma f_i t_i,$$

in which f_i is the frequency of appearance of the ith symbol and t_i is the time required to transmit the symbol. In order to reduce the transmission time to a minimum, the shortest symbols should be assigned to the words with highest frequency.

As an example, consider a possible coding for the letters in the

English language. The probability of occurrence of English letters has been calculated and is given in Table 8-2.

TABLE 8-2. Probability of Occurrence of English Letters

SYMBOL	PROBABILITY, p	SYMBOL	PROBABILITY, p
Word space, or "blank"	0.2	L	0.029
		C	0.023
E	0.105	FU	0.0225
T	0.072	M	0.021
O	0.0654	P	0.0175
A	0.063	YW	0.012
N	0.059	G	0.011
I	0.055	B	0.0105
R	0.054	V	0.008
S	0.052	K	0.003
H	0.047	X	0.002
D	0.035	JQZ	0.001

Source: L. Brillouin, *Science and Information Theory* (New York: Academic Press, 1960).

Using our principle of short transmission times, we would give the longest symbol to the letter that appears least frequently—either *J*, *Q* or *Z*—the shortest to the letter that comes up most frequently—surprisingly, it is a word space, or "blank," because English words are very short.

The question now arises as to what symbols can be used to represent the letters; in principle, any coding method would do. In practice, however, the coding symbols are dictated by the type of hardware used in the real-life coding-decoding machines: the digital computers. These machines utilize components that count or perform logical operations by making binary choices of the type "yes" or "no." In essence, a computer or computerlike circuit consists of a string of boxes such as:

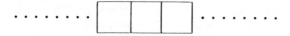

each of which can have a "yes" or a "no" mark inside. The "yes-no's" correspond to the presence or absence of a given voltage in the real computer circuit and are represented by either a 1 or a 0, by convention. At a given instant a string of four "boxes" may look, for example, like:

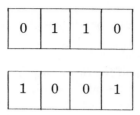

or like

How many different possible ways of marking the boxes with 0's and 1's are there? We would like to find this number because it is clear that each different series or set of numbers could stand for a different letter. In order to figure this out, we could use the branched diagram shown in Fig. 8-5, in which every time we decide whether the box carries a 0 or a 1 there are two new possible alternatives. It is clear that if the system consists of one box only, the choices are 0 and 1 and the system can code only for two letters; when we add another box, the possible choices are 00, 01, 11, 10 as

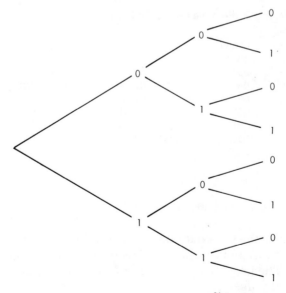

Fig. 8-5. Binary choices give a total number of 2^N possible outcomes for N choices. In this figure $N = 3$ and $2^3 = 8$ choices.

shown in the figure and the system codes for four letters. Similarly, three boxes can code for eight letters: 000, 001, 010, 011, 100, 101, 110, 111. In general, every time we add a new box we multiply the number of choices of *configurations* by a factor of two, so that if the system has H boxes, the number of possible choices is:

$$(8\text{-}1) \qquad\qquad n = 2^H .$$

Thus, for $H = 1$, $n = 2$ and for $H = 3$, $n = 8$. Since there are 27 letters (including the blank, or "space") in the English language, we will need five boxes or choices; with $H = 5$, we obtain for the number of letters that can be coded:

$$n = 2^5 = 2 \times 2 \times 2 \times 2 \times 2 = 32,$$

which is larger than the required twenty-seven letters but necessary because $H = 4$ would give

$$n = 2^4 = 2 \times 2 \times 2 \times 2 = 16$$

which is too small.

A possible code could be that given in Table 8-3, in which we have made sure that each letter is represented by a different set of numbers.

TABLE 8-3. Possible Binary Code for English Letters

blank	00000	H	01001	W	10010
E	00001	D	01010	G	10011
T	00010	L	01011	B	10100
O	00011	C	01100	V	10101
A	00100	F	01101	K	10110
N	00101	U	01110	X	10111
I	00110	M	01111	J	11000
R	00111	P	10000	Q	11001
S	01000	Y	10001	Z	11010

8-6. Probability and Information Content

Let us consider the 0-1 boxes once more and think of each of the $n = 2^H$ possible configurations as the possible outcome of an experiment or message. The probability that any of these messages will appear is, if all are equally likely,

$$(8\text{-}2) \qquad\qquad p = \frac{1}{2^H} = 2^{-H} .$$

This equation can also be written as

(8-3) $$H = -\log_2 p,$$

in which $\log_2$ denotes the logarithm in the base 2. Although the logarithm of a number is a dimensionless quantity, we can keep in mind that H is going to be, in a sense, the number (of boxes) needed to transmit the set of messages. For this reason, H is usually given in terms of binary units—for the yes-no or 0-1 binary decision made at each box—or "bits." H is called the *information content* of the set of messages.

What is the information content, H, of an arbitrary message in the English language if all the letters are transmitted with equal probability? The probability of each letter being transmitted would be:

$$p = \frac{1}{27}$$

so the information content is:

$$H = -\log_2 \frac{1}{27}$$

$$= 4.76 \text{ bits (per letter)},$$

which confirms our previous result that we need more than four binary boxes and less than five to code for such a message. During the process of coding or decoding we cannot use fractional boxes: Since we need more than four, we must use five.

Actually, we have assumed that all the probabilities are equal; if they are not and the probabilities of receiving messages 1, 2, 3, etc., are p_1, p_2, p_3, etc., respectively, the information content is *defined* as:

(8-4) $$H = -p_1 \log_2 p_1 - p_2 \log p_2 - p_3 \log p_3 - \ldots.$$

This formula reduces to the previous expression (Eq. 8-3) when all the probabilities are equal. The actual frequencies of the various English letters was given in Table 8-1; the information content calculated this way is:

$$H = -0.2 \log_2 0.2 - 0.105 \log_2 0105 - \ldots$$

$$= 4.03 \text{ bits.}$$

Since the information content is less than for the case of equal probabilities, the number of binary decisions needed is slightly smaller.

The information content decreases still further if we include the fact that the probability of appearance of a given letter in an English word is also determined by the letter that precedes it. The probability that an *"h"* will follow a *"t"* is great, for example; this phenomenon is called *redundancy*. Under conditions in which we consider two letters at a time, the information content H reduced to $H = 3.32$. If we take three letters at a time, the information reduces to $H = 3.1$.

This simple example illustrates two important points of information theory:

- H is largest when all probabilities are equal.
- H decreases when the probability of a given event is conditioned to the probability of appearance for another message.

8.7 Encoding and Decoding Redundant Messages

In common usage, the word redundant implies that parts of the information are not needed to recover the original message. Shannon's redundancy, the technical concept introduced in information theory, attempts to give a measure of everyday redundancy. Thus, the redundancy of the English language is 50% when entire words are considered, which implies that on the average half of the words in an English text can be crossed out without losing the sense. There is also a high redundancy in the English alphabet itself since, as we saw in Sec. 8-6, the presence or absence of a given letter is strongly dependent on the previous letters in the message. Unfortunately, the calculated redundancy cannot always be equated with the intuitive term. A good example of how misleading this type of calculation can be is provided by redundancy in the genetic code.

The redundancy of a given DNA molecule can be estimated by analyzing the constraints imposed on the probability of presence of neighboring sequences. If information is available about the frequencies of the four DNA bases, the amount of redundancy can be found. Lila Gatlin has calculated the redundancy of various DNA's and found that it varies between 0 and 11%—is rather low. Some DNA's, however, show high redundancy. A special DNA, the satellite DNA of the crab, has 83% redundancy, and an RNA associated with reovirus has the composition

$$A = 87.8\%$$

$$U = 10.5\%$$

$$C = 1.4\%$$

$$G = 0.3\%$$

which makes the genetic information highly redundant (see Prob. 8-3). But the question arises as to what the genetic redundancy measures. If a word is crossed out of an English text, we can recover the message most of the time. The same occurs if a word appears in the wrong position. If a base is deleted or changed in a DNA molecule, however, the results are disastrous most of the time. Consider an example in which the sequence

-A-A-G-G-$\underline{G}$-U-C-C-A-U-C-A-C-U-U-A-A

is transformed, by deletion of the underlined base to the sequence

-A-A-G-G-U-C-C-A-U-C-A-C-U-U-A-A.

This is a mutation which actually occurs in the lysozyme of bacteriophage T4 (a bacterial virus) and which effects the change from

-Lys-Ser-Pro-Ser-Leu-Asp-Ala-

to   -Lys-Val-His-Leu-Met-Ala-.

While in this specific example, the reading in triplets seems to eventually get back "in frame," such changes usually give proteins with completely different properties. One of the most striking examples is provided by the amino acid mutations of hemoglobin, which occur by replacement of single DNA bases.

8-8. Biological Mistakes

From time to time an enzyme will make a mistake, recognizing the wrong substrate, but in general the error will not be more serious than saying "hello" to someone who looked familiar but we did not really know. In a few cases, the lack of absolute knowledge about the position and chemical nature of every atom can lead to serious physiological mistakes. As an example, consider sickle cell anemia, a disease common in certain parts of the world, in which red blood cells adopt deformed shapes when the oxygen in the tissues falls below a certain level. When the structure of hemoglobin in the sickle cell is studied, it is found that the only difference between it and normal hemoglobin is the change in a single amino acid, as follows:

Normal HB: Val-His-Leu-Thr-Pro-Glu-Glu . . .

Sickle cell Hb: Val-His-Leu-Thr-Pro-Val-Glu . . .

Calculation of the *genetic* redundancy, then, does not provide the type of information the theory set out to measure: Deletion of a single "word" is enough to knock out the whole message. This is partly because the genetic code does not constitute the final message, and information theory deals with the final message at the output. These are the *functional* genetic messages as observed in the form of a specific chemical step, programmed behavior, etc.

8.9. Error-correcting Codes

How can errors be prevented or at least reduced? The design of error correcting or error detecting codes depends on the type of system and on the ingenuity of the encoder. Whatever the system, we always pay a price to insure that the message will be correctly received. The main *cost* is the complexity of the transmitter. This is a "fixed" cost. The main *disadvantage* is the time delay required to transmit an absolutely correct message.

Consider the example of telegram transmission. Anybody who has sent or received a telegram knows how easily errors can be transmitted. One way to make sure that the message has been correctly sent is to ask the receiving end to retransmit the message back. If the retransmitted message looks exactly like the message originally sent, we may *suspect* that the original transmission was accurate, but we cannot be absolutely certain, because the two message could have contained mistakes that canceled out. The only way to make certain is to keep retransmitting the message . . . forever. Of course, every time we get the right message back we become more confident; the possibility of an error decreases. If one word out of ten, on the average, is wrong every time the message is transmitted the expected error will be

0.10 (or 10%) after the first transmission

0.10 X 0.10 = 0.01 (1%) after the second transmission

0.01 X 0.10 = 0.001 (0.1%) after the third transmission

0.001 X 0.10 = 0.0001 (0.01%) after the fourth transmission

and so on. Clearly, this process is similar to a reversible process: If we are willing to send the message "infinitely" slowly, we can approximate the perfect transmission as closely as we wish.

The "very safe" code usually takes an infinite time to transmit and decode. A possible way to get around this problem is by allowing some mistakes in the individual words being decoded while providing some real redundancy in the overall message. In this way, the whole paragraph will make sense even though there may be mistakes in the individual spellings. Another way consists in giving check marks at different steps in the transmission process. In this way, errors are not carried through to the next step. This is the method used to assemble proteins.

A specific example is given by the errors made in "loading" a given transfer RNA with the correct amino acids. As the t-RNA molecule acts as the "adaptor" between a given codon and the specific amino acid, the step is critical in protein synthesis. The process consists of two steps: the activation of the amino acid to form the high energy Amino acyl-AMP and the actual bonding of the amino acid to the t-RNA. Table 8-4 gives the recognition errors made by the enzymes specific for a given amino acid. Clearly, each enzyme always loads a certain amount of the wrong amino acid, but the fraction of wrong amino-acyl SRNA complex finally formed is small, usually negligible.

TABLE 8-4. Errors Made When Loading a Specific t-RNA with the Correct Amino Acid

ENZYME SPECIFIC FOR	AMINO ACID USED	A ACYL AMP FORMATION	A A-tRNA FORMED
Leucine	Leucine	358	3.2
	Valine	13.2	0.18
	Methionine	<8.0	<0.01
	Isoleucine	<3.0	<0.01
Valine	Valine	560	25
	Leucine	<0.5	<0.01
	Isoleucine	<4.0	<0.01
	Methionine	<2.0	<0.01
Isoleucine	Isoleucine	768	3.3
	Leucine	31	0.07
	Valine	416	<0.03
	Methionine	41	0.07

Source: V. Ingram, *The Biosynthesis of Macromolecules.* (New York: W. A. Benjamin, Inc., 1965).

8-10. Information Theory and Thermodynamics; Maxwell's Demon

We said that a gas at rest actually consists of many molecules moving in every direction. Although any large sample of these molecules will possess the same average velocity, some molecules are faster than others. When physicist James Clark Maxwell published his book *The Theory of Heat* in 1871, he thought it was possible to violate the Second Law of Thermodynamics by separating the fast from the slow molecules:

> Now let us suppose that (a vessel full of air) is divided into two portions, A and B, by a division in which there is a small hole, and that a being, who can see the individual molecules, opens and closes this hole, so as to allow only the swifter molecules to pass from A to B, and only the slower ones to pass from A to A. He will thus, without the expenditure of work raise the temperature of B and lower the temperature of A, in contradiction to the Second Law of Thermodynamics.

This hypothetical being is called *the Maxwell demon.* It took about a hundred years to show why Maxwell's argument is a fallacy: Any biological system, demon or not, has an entropy increase that is larger than the decrease in entropy produced by it.

Physicists were at first so convinced that the animal body did not act as a thermodynamic engine that Lord Kelvin made sure he included the phrase "inanimate material agency" in his original definition of the Second Law (1851), and added:

> The means in the animal body by which mechanical efforts are produced cannot be arrived at without more experiment and observation . . . . Whatever the nature of these means, consciousness teaches every individual that they are to some extent, subject to the direction of his will . . . . The conception of a sorting demon is purely mechanical . . . . It was not invented to help us deal with questions regarding the influence of life and of mind on the notions of matter, questions beyond the range of mere dynamics.

Many physicists worked on the problem of the demon for years before Leon Brillouin, in 1951, showed that any intelligent being—any entity able to store information—has to increase its own entropy before it can effect an environmental reduction in the entropy of the external world. Brillouin first showed, using kinetic reasoning, that the demon would not be able to see individual molecules but would instead detect an average energy. This would make the inside of the box a "black body," and in order to see the molecules the demon would need a light, or "torch." This torch would provide the energy required to distinguish the fast from the slow molecules. Furthermore, Brillouin concluded that the same type of process was actually going on when an experimenter makes a measurement in the laboratory; in order to obtain a number (symbols), he must provide a certain amount of energy (by plugging the measuring instrument in an electrical outlet, for example) or draw some energy from the system he is trying to describe. These examples—the demon and the scientist—indicate that there is a close relationship between sorting out symbols and energetic processes.

8-11. Information Content and Entropy

Equation 8-3 is clearly related to the statistical formulation of entropy. There is nothing in information theory that prevents us from considering the configurations, or states, in which a given physical system can be found as "messages." Some of these messages are highly probable, as we have seen in Chapter 6, and some are less probable. If we consider the simple case in which all states (messages) have equal probability, the entropy is given by

$$S = k \ \ln p.$$

We can write $p = 2^{-H}$, as before, to obtain:

$$(8\text{-}5) \qquad\qquad S = k \ \ln \frac{1}{2^H}$$

$$= k \ln 1 - k \ln 2^{-H}$$

Since the logarithm of 1 is 0, the expression reduces to:

$$(8\text{-}6) \qquad\qquad S = -k \ \ln 2^{-H}$$

Furthermore, we can utilize the property that $\log a^b = b \log a$ to obtain:

$$(8\text{-}7) \qquad\qquad S = Hk \ln 2$$

Information and entropy, then, are closely related to one another.

8-12. Reversible vs. Irreversible Transmitters

As there is a tendency for information to decrease, we should expect that under general conditions of spontaneous processes some information will be lost—in the same sense that some entropy is

TABLE 8-5. Energy Required in Various Human Communication Activities

ACTIVITY	ENERGY (joules)	INFORMATION CONTENT (bits)	ENERGY PER INFORMATION (joules per bit)
Audio record activities:			
Telephone conversation (1 min)	2,400	288,000	.008
High fidelity audio record playback (1 min)	3,000	2,400,000	.001
AM radio broadcast (1 min)	600	1,200,000	.0005
Pictorial record activities:			
Telecopy 1 page (telephone facsimile)	20,000	576,000	.03
Projection of 35 mm slide (1 min)	30,000	2,000,000	.02
Copy one page (xerographic copy)	1,500	1,000,000	.002
Print 1 high-quality opaque photographic print (5″ × 7″)	10,000	50,000,000	.0002
Project 1 television frame (1/30 sec)	6	300,000	.00002
Character record activities:			
Type 1 page (electric typewriter)	30,000	21,000	1.4
Read 1 page (energy of illumination)	5,400	21,000	.3
Copy 1 page (xerographic copy)	1,500	21,000	.07
Print 1 page of computer output (60 lines × 120 characters)	1,500	50,400	.03

Source: M. Tribus and E. C. Irvine, "Energy and Information," *Scientific American*, September 1971.

"gained." In fact, we can define a *reversible* transmitter as one that maintains the information content constant and an irreversible transmitter as one in which the information decreases as the message is transmitted.

Symbol converters do not increase in entropy—decrease information—by the same order of magnitude as regular energy transducers. This principle has been somewhat irreverently stated as: "Talk is cheaper than action." Little energy is required to manipulate information. It is for this reason that our civilization can handle large amounts of energy by utilizing devices that control rather than generate or transform energy. The fact that Eq. 8-7 took so long to be detected should indicate that the energy required to organize information is, in general, very low. Table 8-5 shows typical energy expenditure for recording activities.

8-13. Information and Knowledge

The choice of the term information theory is somewhat misleading. While the word information has connotations of "meaning" and "knowledge," information theory deals merely with the *statistics* of message transmission. Probability, not meaning, is the leitmotif of information theory. A spy story and a philosophical work will have the same information content if the probability of appearance of words is the same in both. Furthermore, classical information theory does not include the relative importance of messages: "Happy birthday," "you are sick," and "there will be an earthquake" are assigned the same value in the context of this model.

Problems

8-1. Show that the information content of a message in English is $M = 4.76$ if all the letters are assumed to appear with the same frequency.

8-2. Calculate the information content of a genetic message in which all nucleotide sequences are equally frequent and the final message is read in terms of the amino acid frequency (refer to the genetic code table). Repeat, assuming all amino acids are equally likely to occur. Discuss.

8-3. The nucleotide composition of reovirus DNA is A = 87.8%, U = 10.5%, C = 1.4%, G = 0.3%. Calculate the *a priori* percentages of each of the triplets and use these frequencies to find the information content.

8-4. What is the minimum information content of a "Maxwell demon" which establishes a concentration ratio of $c_1/c_2 = 2$ molar at a temperature of $25°$ C? Assume that the system starts as two compartments containing 1 liter of 1 molar solution each. Suppose that the demon is an enzyme in which all amino acids appear with equal probability. What is the minimum number of amino acids such an enzyme would have?

8-5. Calculate the percent error made by the *leucine* specific amino acid activating enzyme in recognizing *valine*. Calculate the error made in the second step alone. Comment on the possible advantage of having sequential, rather than parallel, molecular assembly.

9

INFORMATION THEORY
AND BIOLOGY

One of the interesting aspects of common biological messages is that little information is required to encode them. This is especially surprising in view of the large number of possible *genetic* combinations. The small number of *functional*, or phenotypic, messages clearly indicates that *most genetic bits effectively carry no information.* A few relevant examples will illustrate this point.

9-1. How Many Bits Can the Human Brain Store?

We all know people who can remember long lists of names, birthday dates, etc., and some who cannot remember their own name. Furthermore, the ability to remember long strings of symbols becomes more obvious in cases where the span of *immediate* memory is being tested. It looks, then, as if some people have the ability to store many more bits of information than can others.

As it turns out, the important quantity is *the number of items* to be remembered. Psychologist George A. Miller was the first to point this out in his article "The Magical Number Seven, Plus or Minus Two," though the basic idea had been expressed by Descartes.

Most people can remember instantaneously about seven bits—they can code for that many binary choices and retrieve the information immediately. The actual number of bits contained in this information depends only on how clever they are at coding. Miller indicates that the efficient way of coding is in "informationally rich units." We must keep words instead of letters, and sentences instead of words. The resultant information is greatly increased by storing symbols this way.

Consider, for example, three possible choices for keeping information: in decimal digits, alphabetic characters (letters), or monosyllables. The information contained in symbols taken from each of these groups is:

(1) $\log_2 10 = 3.3$,

(2) $\log_2 26 = 4.7$, and

(3) $\log_2 1000 \cong 10$,*

respectively. The total information kept when remembering five symbols is 16.5, 23.5, and 50 for the numbers, letters, and monosyllables, respectively. But even though the amount of information is different, five symbols, whether consisting of numbers, letters, or monosyllables, are equally difficult to re-member!

9-2. Is Biological Memory Stored in Informational Macromolecules?

Some recent experiments on goldfish seem to suggest that the origin of biological memory may be related to the availability of informational macromolecules DNA or RNA, and the translation of this information into structural or catalytic proteins.

When goldfish are trained to move between the two compart-ments of a divided swimming pool, the fish learn to do the easy task without difficulty. However, if an injection of puromycin is given to the fish's brain immediately after training, all memory of training is erased and the goldfish never learns the task (Fig. 9-1). Since injection of puromycin 1 hour after training causes no such memory loss, it looks as if the antibiotic blocks some process that takes place in less than an hour. Bernard Agranoff speculates that the affected process is the release of the nascent protein from the ribosome, and concludes that long-term memory is directly related to the synthesis of new proteins (Fig. 9-2). This theory is not endorsed by every physiologist, psychologist, and molecular biologist, because it is hard to know what the primary effect of the injection is: Puromycin blocks synthesis of several proteins, and it also puts the fish in a state of shock. Nevertheless, the idea that memory is related to nucleic acid information storage has been the subject of several recent

*There are about 1,000 common monosyllables in the English language.

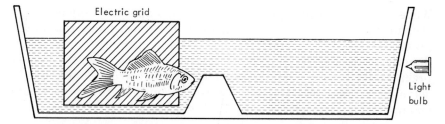

Fig. 9-1. Principle of training tank utilized by Agranoff to teach gold-fish to respond to light; electric shocks were applied through the electrical grids at the same time that the light was turned on. Eventually, the fish learned to move to the lighted compartment without application of a shock. (B. W. Agranoff, "Brain Correlates of Learning" in *The Neurosciences*, ed. by G. C. Quarton et al. New York: Rockefeller University Press, 1967.)

theories. It has been speculated, for example, that the physico-chemical basis of memory lies in the enzymatic modification of DNA in nerve cells, which in turn controls the *rate* of protein synthesis, rather than in the kinds of proteins made. Other theories propose that the brain stores information in holograms similar to the two-dimensional photographs containing three-dimensional information, or even that memory is regulated by the rate at which cells die.

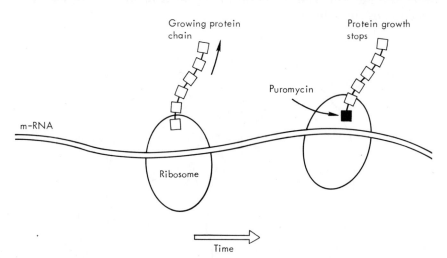

Fig. 9-2. Protein-blocking agents such as puromycin block formation of the nascent protein chain at the ribosome. Thus, the fact that learning of a simple task can be prevented by injection of puromycin—to the goldfish in Fig. 9-1, for example—could indicate that short-term memory is mediated by protein synthesis. (See also B. W. Agranoff, "Memory and Protein Synthesis," *Scientific American*, June 1967.)

While it is not possible to prove or disprove experimentally and with absolute certainty *any* of these theories, it is most likely that in the present stage of evolution of living systems, memory is not coded directly in nucleotide sequences. If that were the case, the human brain would have a memory capacity of 10^{16} to 10^{20} bits, but in actuality a brain that acquires 3 bits a second for 100 years would end up having a capacity of only 10^{10} bits!

9-3. Dancing Bees and Chemical Ants

The small number of genetic messages required to program some important activities of lower animals is well demonstrated in the programmed social behavior of some insects.

Karl von Frisch, one of the most dedicated students of the social activities of bees, discovered that the dances of these insects indicate the position of nectar relative to the hive. The dwarf honey bee, for example, dances in a horizontal plane roughly describing the shape of a figure *8* when nectar is available between 50 and 1000 feet away; in order to indicate the distance from the hive to the source of nectar, the bee waggles her abdomen (Fig. 9-3)—the closer the source, the faster the motion. The direction of the wagging dance indicates the direction of the source.

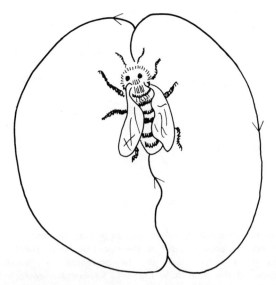

Fig. 9-3. Wagging dance of the honeybee indicates direction and distance from the hive to the food source.

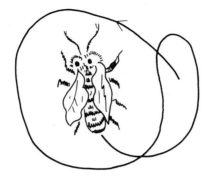

Fig. 9-4. Round dance of the honeybee is performed when food is very close to the hive.

If the food is very close to the hive, the bee performs a different dance, called round dance (Fig. 9-4). It turns out, however, that bees receiving the signal can detect only four different ranges centered around 10, 50, 200, and 1000 feet—corresponding to 0, 9.8, 7, and 4 waggles. The information content of these equiprobable messages is, then,

$$H = -\ln_2 p = -\ln_2 \frac{1}{4}$$

$$= \ln_2 4,$$

or, $\qquad\qquad H = 2$ bits.

In the case of other species, there seem to be five, rather than four, messages (Fig. 9-5). The information content is $H = -\ln_2 (1/5) = \ln_2 5$. We can conclude, then, that distance information contains about 2 bits. Direction information, on the other hand, contains 4 bits; that is, a bee placed at the center of a circle can send distinct messages in 16 possible directions, an accuracy of 22.5 degrees.

Another instance of programmed behavior is found in fire ant pheromone communication. Throughout the plant and animal kingdoms, there are substances that can send signals to distant organs of the same individual, thus affecting the rate of a specific metabolic activity. These substances are called *hormones.* There is another family of substances which are similar to hormones but which act on the external world influencing the behavior of a different individual of the same species; these are *pheromones* and they have been extensively studied in ants by Edward Wilson and his colleagues at Harvard University.

There are different types of pheromones for various activities— sex, eating, fear—but the fire ant worker's trail involved in the indication of distance and direction of food or nest deserves

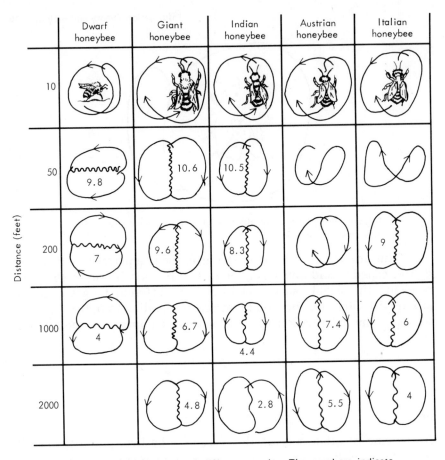

	Dwarf honeybee	Giant honeybee	Indian honeybee	Austrian honeybee	Italian honeybee
10					
50	9.8	10.6	10.5		
200	7	9.6	8.3		9
1000	4	6.7	4.4	7.4	6
2000		4.8	2.8	5.5	4

Distance (feet)

Fig. 9-5. Dances in bees of different species. The numbers indicate the number of wagging motions done in 15 seconds. (A detailed description may be found in K. von Frisch, "Dialects in the Language of Bees," *Scientific American*, August 1962.)

particular attention, because it is the best understood from the informational point of view.

The fire ant leaves a rather narrow trail of pheromone (Fig. 9-6) which can be perceived by other ants. The fast rate of evaporation guarantees that the signals will be perceived only when there is food available. According to Wilson, the trail is reinforced by the pheromone secretion of worker ants who return to the nest after feeding.

The distance-information content is about the same as in the dwarf bee, 2 bits (4 possibilities), although the mechanisms in the fire ant and in the dwarf bee are not related in any evolutionary way. Directional information in the ant has a peculiarity that arises from the shape of the ant's trail: Information content changes with the

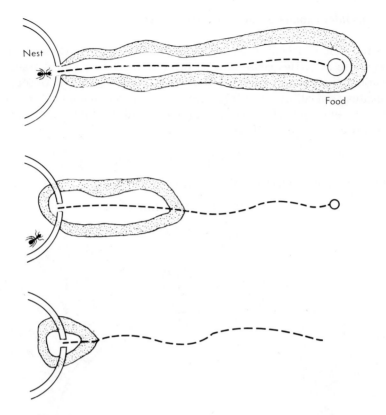

Fig. 9-6. Disappearance of the working fire ant trail in time assures that no pheromone is present when food has been exhausted. (E. O. Wilson, "Pheromones," *Scientific American*, W. J. Freeman and Co: San Francisco, May 1963.)

distance of reception. Thus, at 2.5 centimeters, direction can be given with a precision of about 45 degrees, which means that there are 8 equiprobable directions. Accuracy increases with distance, however, and at 10 centimeters there are 32 equiprobable directions (Table 9-1).

TABLE 9-1.

DISTANCE (cm)	DIRECTIONS (number of distinct messages)	H (bits)
0–2.5	8	$\log_2 8 = 3$
2.5–5	16	$\log_2 16 = 4$
5–7.5	24	$\log_2 24 = 4.6$
7.5–10	32	$\log_2 32 = 5$

9-4. The Information Content of a Bacterial Cell

It is possible to freeze a bacterium down to temperatures in which all enzymatic activities cease and to bring it back to room temperature without significant loss of biological activity. This observation indicates that all of the information needed to make a bacterium like absolute entropy of a bacterium measured in this experiment is, according to physicist H. Quastler,

$$S = 9.3 \times 10^{-12} \text{ cal/deg},$$

which gives an information content of

$$H = 4 \times 10^{12} \text{ bits.}$$

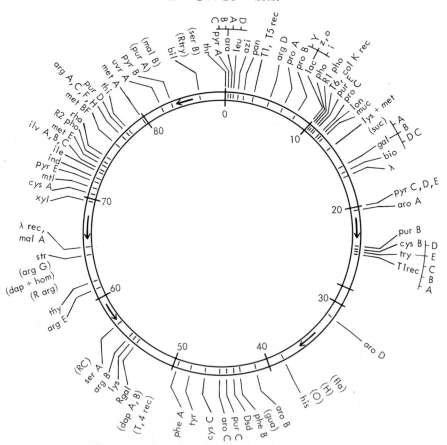

Fig. 9-7. Genetic map of *E. coli* showing some of the known mutations of the bacterium. (J. Watson, *The Molecular Biology of the Gene.* New York: W. A. Benjamin, Inc., 1965.)

The genetic variability of a bacterium, however, is much smaller, at most in the order of 1,000 binary choices. This discrepancy is again found in the calculation of the information content needed for an enzyme to recognize a substrate if the enzyme were to do the job in a perfect way, recognizing each and every atom. Operationally, the enzyme does not recognize each individual atom but only small groups of atoms arranged in certain recognizable shapes, in the same way that we recognize a friend using a few characteristic features. Living systems, then, do not approach the problem in the way the analytical biochemist would: they do not have to know exact formulas, only general properties and shapes.

9-5. Functional Number of Genetic Messages

The genetic map of *E. coli*, however, reveals fewer than 1,000 locations for possible mutations, as shown in Fig. 9-7, so from the functional point of view, the upper limit can at most be 1,000. A lower bound is estimated by evolutionary biologist Conrad Waddington at 106; he simply calculates how many enzymes any cell must have to perform its essential jobs.

These enzymes are the ones that take part in the following metabolic processes:*

glycolysis	15
citric acid cycle	12
β-oxidation of fatty acids	5
Hexose monophosphate shunt	13
Oxidative phosphorilation	4
Fatty acid synthesis	4
Pyrimidine synthesis	10
Purine synthesis	15
Amino acid activating enzymes	20
Methylation and one carbon metabolism	8
	106

Waddington also points out that when different DNA's are compared from an evolutionary point of view, the dissimilarities between closely related species are not as large as the similarities. Furthermore, the external characteristics of different plants and animals are not as extreme as the casual observer could be led to believe. In the

*From Conrad H. Waddington, "Biological Organization and Physical Systems," in *Biology and the Physical Sciences* edited by Samuel Devons (New York: Columbia University Press, 1969).

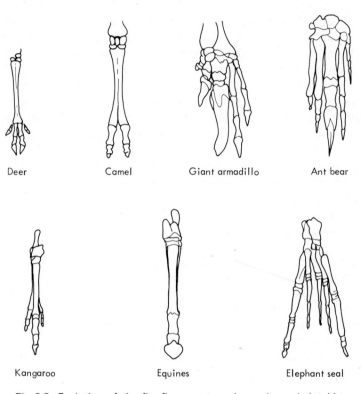

Fig. 9-8. Evolution of the five-finger pattern shows close relationship in the feet of various animals. (C. Waddington, "Biological Organization and Physical Systems," in *Biology and the Physical Sciences*, ed. by S. Devons. New York: Columbia University Press, 1969, adapted from P. Tschumi, *Rev. Suisse Zoologie.*)

evolution of the five-finger pattern in some animals, for example, it is seen that the changes can be achieved by a digit becoming longer or shorter or deleted, although the change in one digit affects the global—or overall—characteristics of the foot (Fig. 9-8).

Similar examples of small variations affecting the global pattern are found in inorganic crystals. As an example, we can consider NaCl, one of the most readily available; its molecular structure is simple, being organized—at the atomic level—in cubes of atomic dimensions. At the microscopic level, this structure manifests itself also in the form of a cube. In the presence of urea, however, the corners of the cube are sliced, as shown in Fig. 9-9. If enough urea is present, the corner surfaces completely dominate the shape of the crystal. Similarly, a small change in a large molecule such as a protein should be able to produce drastic changes in structure and function.

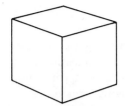  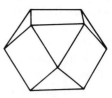

Fig. 9-9. The addition of small amounts of urea can lead to drastic changes in the shape of a simple crystal such as the cubic NaCl. This suggests how minimal changes in the conformation of a large protein molecule may lead to striking macroscopic changes in shape.

9-6. The Emergence of Life as an Accident

Quastler used the statistical thermodynamic approach to estimate the probability that a 1,000-bit life (e.g., bacterium) emerged as a creative accident in which the right atoms arrived in the right order to a certain position on earth. The number of possible bacteria that could have been formed is given by the ratio of the total volume available—the oceans of the earth—to the volume of a single bacterium. This gives an upper number of 5×10^{32} sites if the whole earth covered with water was available for the reaction, and a lower number of 5×10^{10} sites if only certain regions close to the shores were used.

The upper number giving the time of emergence of life is

$$2 \times 10^9 \text{ years} = 2 \times 10^{13} \text{ hours.}$$

The minimum time of assembly is the time it takes for a bacterium to do the job by itself: one hour. The maximum time needed is in the order of 10^{10} hours, when no enzymatic activities are present. In the best of all possible worlds, then, it would be possible to make 2×10^{13} bacteria in 2×10^9 years. Multiplying this number by the maximum number of sites, we obtain,

$$(2 \times 10^{13}) (5 \times 10^{32}) = 10^{46}$$

possibilities of emergence for each message. Thus, the probability of making a 1,000-bit life in this way is expressed by

$$p = \frac{10^{46}}{2^{1000}} = 10^{-254}$$

or 1 divided by 10 followed by 254 zeroes! Clearly, this is a very unlikely event, so it is almost certain that life on earth did not begin in this way.

9-7. How Did Life Begin?

We have shown that it is unlikely that the beginning of life took place in a sudden fashion at a local place on earth; the question arises as to whether it is possible that some other kind of spontaneous physical mechanism will explain the origin of living organisms. We should point out, first, that the relative abundance of the first 31 elements in the cosmos is closely correlated with the relative abundance of the same elements in the biosphere (Fig. 9-10); this observation alone should lead us to suspect that life is a phenomenon which is related to the physical laws of the universe.

What do physical laws tell us about the possible origin of life? The most likely model is suggested by Aleksandr Oparin: that

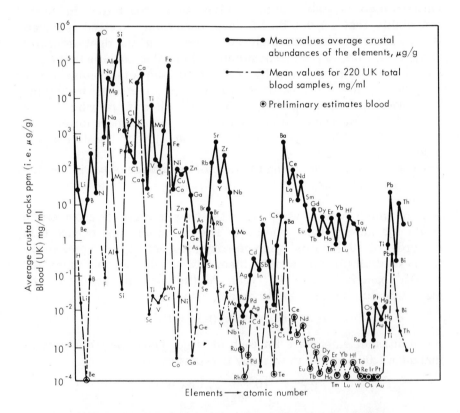

Fig. 9-10. Relative abundance of elements in human blood and the earth's crust showing close correlation between the two. (J. Tinker, "Measuring the Elements in Man," *New Scientist*, September 1971.)

molecules were formed in a sequential—as opposed to a simultane-
ous—way; simple molecules being made first, then coming together,
after long periods of time to form larger, more complex molecules.
Since the original atmosphere of the earth was a reducing rather than
an oxidizing one, the first molecules "survived" (that is, remained
stable) in the primitive ocean—CH_4, NH_3, H_2O, H_2—for long periods
of time. The large amounts of energy that could then reach the
earth's surface provide the necessary overall increase in entropy for
the reactions to take place (otherwise, the formation of more and
more complex molecules would imply a violation of the Second Law
of Thermodynamics).

Given the distribution of the original oceanic compounds
(ammonia, water, methane, and hydrogen) in the equilibrium
situation, we now ask for the probability of forming a molecule like
ethanol, CH_3-CH_2OH. Clearly, this is an unlikely event, as we are
asking for molecules that are comfortably resting in their equilibrium
states to split and then form a different molecule that is not in
equilibrium with the rest of the sea. Spontaneously, the process has
little chance of occurring. But again, how little is little? An
improbable process will always occur if: (a) we wait long enough or
(b) we perform the experiment many times. Biophysicist Harold
Morowitz has calculated the thermodynamic probabilities expected
using the entropic methods we discussed in Chapter 4. Starting with
an atmosphere of CO_2, N_2, and H_2O, he calculates the following
relative compositions for different compounds:

Water (H_2O)	2.24	Ethane (CH_3-CH_3)	2.7×10^{-8}
Carbon dioxide (CO_2)	0.88	Acetic acid	
Nitrogen (N_2)	0.50	(CH_3-COOH)	0.029×10^{-8}
Methane (CH_4)	0.12	Pyruvic acid	
Hydrogen (H_2)	0.027	(CH_3COCOOH)	0.42×10^{-8}
Ammonia (NH_3)	0.000054		

It is possible to perform a real experiment in which the original
chemical conditions of the earth are reproduced and the molecules
obtained are observed. Harold Urey and Stanley Miller used the flask
shown in Fig. 9-11 with a mixture of CH_4, NH_3, H_2O, and H_2,
which was made to react while energy was provided in the form of
electrical sparks. The yields of different molecules are given in Table
9-2.

TABLE 9-2. Yields from Sparking a Mixture
of CH_4, NH_3, H_2O, and H_2

COMPOUND	YIELD [moles (X 10^5)]
Glycine	63.
Glycolic acid	56.
Sarcosine	5.
Alanine	34.
Lactic acid	31.
N-Methylalanine	1.
α-Amino-n-butyric acid	5.
α-Aminoisobutyric acid	0.1
α-Hydroxybutyric acid	5.
β—Alanine	15.
Succinic acid	4.
Aspartic acid	0.4
Glutamic acid	0.6
Iminodiacetic acid	5.5
Iminoacetic-propionic acid	1.5
Formic acid	233.
Acetic acid	15.
Propionic acid	13.
Urea	2.0
N-Methyl urea	1.5

Source: S. Miller, "The Origin of Life," in
Biology and the Physical Sciences ed. by
S. Devons (New York: Columbia University Press, 1969).

It is seen that the results of the experiment agree with the thermodynamic calculation. We can conclude, then, that Oparin's hypothesis is substantiated by both experimental and theoretical evidence. Once the essential molecules of life were formed, energy metabolism of various forms of life probably evolved following the scheme proposed by biologist George Wald (Fig. 9-12).

Once again, we must stress that entropy did not decrease because of the organization taking place on earth: Large amounts of energy were degraded and there was an overall increase, rather than a decrease, in entropy.

9-8. Life in Other Worlds?

Is it possible that life arose on other planets in the same way that it appeared on earth? First, it is unlikely that life of the type we know

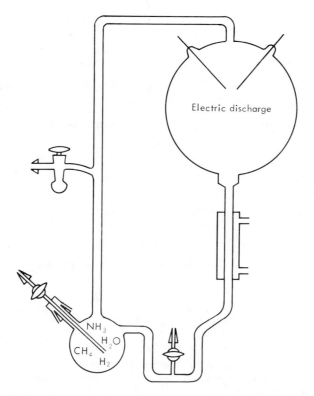

Fig. 9-11. Apparatus used by Miller to simulate the original chemical reactions of life on earth. Voltage sparks or ultraviolet light is used to provide energy for unlikely chemical reactions to occur in reasonable time.

exists elsewhere in the solar system. There are speculations, however, that life may be a phenomenon of frequent occurrence in the universe, Miller quotes astronomer Harlow Shapley's estimate of the number of stars similar to our sun (more than 10^{20}) in the universe. Of these, Shapley claims, at least a million must have planets like the earth, with atmosphere and energetic conditions similar to those of our planet. The small probability for the appearance of life would reduce the number further, but the probability is still high enough to ascertain that there is almost absolute certainty that life must be present elsewhere in the universe (even if we are never to detect it). It is interesting to point out that there are a number of scientists who are trying to figure out kinds of codes that will serve as interplanetary signals between different civilizations. The rate at which information will be exchanged must necessarily be slow

ANAEROBIC PHASE

1. Fermentation: a chemical source of energy; by-product CO_2
   e.g., $C_6H_{12}O_6 \longrightarrow 2C_2H_5OH + 2CO_2 + 2 \sim P$

2. Hexosemonophosphate cycle: metabolic hydrogen for reductions
   $6C_6H_{12}O_6 + 6H_2O + 12 \sim P \longrightarrow 12H_2 + {}_5C_6H_{12}O_6 + 6CO_2$

3. Photophosphorylation: light into high-energy phosphates
   $$\xrightarrow[\text{chlorophylls, cytochromes}]{\text{light}} \sim P$$

4. Photosynthesis: light into new organic molecules; by-product O_2
   Bacteria:

   $$6CO_2 + 12H_2A \xrightarrow[\text{chlorophyll}]{\text{light}} C_6H_{12}O_6 + 6H_2A + 12A$$

   Algae, higher plants:

   $$6CO_2 + 12H_2O \xrightarrow[\text{chlorophyll}]{\text{light}} C_6H_{12}O_6 + 6H_2O + 6O_2$$

AEROBIC PHASE

5. Respiration: metabolic energy from combustions
   $C_6H_{12}O_6 + 6H_2O + 6O_2 \longrightarrow 6CO_2 + 12H_2O + 30\text{--}40 \sim P$

Fig. 9-12. A possible sequence for the evolution of energy metabolism. (From George Wald, "Radiation and Life," in *Recent Progress in Photobiology*, 1965, p. 333.)

because of the great amount of time required to transmit a signal, even to the closest star.

The possible levels of civilization are classified according to the energy that a given culture (planet) can control. The first step is control of the energy of the planet; the second, control of the energy of the star; and the third, control of the energy of the galaxy. The earth is, of course, only in the first stage.

9-9. Evolution of Proteins

The problem of evolution presents another finite combinatorial challenge which, in principle, can be treated in thermodynamic and informational terms. It turns out, however, that it is exceedingly difficult to quantitate evolutionary changes even though the external description of evolutionary characteristics in plants and animals is relatively simple. As an example of the methods, we may consider

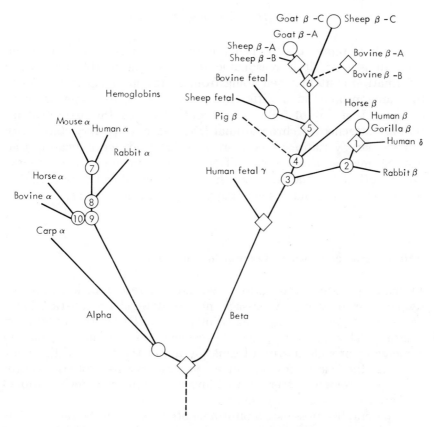

Fig. 9-13. A philogenetic tree giving the evolutionary distance between various species can be constructed by studying differences and similarities in the amino acid composition of a protein such as hemoglobin. (M. O. Dayhoff and R. V. Eck, *Atlas of Protein Sequence and Structure*, National Biomedical Research Foundation, 1969.)

the evolutionary changes in the amino acid sequence of the protein hemoglobin. Thanks to the modern methods of computer protein sequencing and the dedicated work of several laboratories, it is now possible to compare the amino acid changes that have taken place in this protein. The "tree" of Fig. 9-13 gives the distance—in terms of the number of different amino acids—between species. The knowledge of the sequence variation permits us to calculate the change in information content when going from one species to a closely related one. This information change gives the *minimum* entropy needed to effect the evolutionary step; the numerical value is given by the formula,

$$H = -\Sigma p_i \log_2 p_i$$

in which p_i represents the probability of a change in a given amino acid. In general, it is reasonable to expect noticeable changes in information content when going from a protein in a given species to the same protein in a different species. There are some proteins, however, that do not follow this trend: Histones, the basic proteins that are attached to chromosomal DNA, seem to be very stable with evolutionary variations. As there is some evidence indicating that histones may act as a control in DNA replication and in turning genes on and off, it is speculated that the mechanism of cell differentiation has not varied too much since it appeared in living systems.

9-10. Capacity of a Channel: Crayfish Photoreceptor

In many practical applications, we are interested in knowing the dynamic properties of a message, not its static characteristics. Thus, we want to find out how many bits *per unit time* the system can transmit. This number, called the capacity of the channel, may in some cases provide useful information about the nature of the code. This is the case, for example, in the crayfish photoreceptor walking-movement system, a well-known invertebrate device studied by Lawrence Stark and associates.

The crayfish possesses a photoreceptor system in the tail which is connected to the brain by means of two neural axons. The output of the system is a tension on the leg muscles, so that the crayfish moves whenever light falls on its tail. It is experimentally found that the frequency of firing varies between 10 pulses per second—which corresponds to a separation between pulses of 100 milliseconds, and 100 pulses per second—corresponding to a period of 10 milliseconds between pulses.

Stark considers two possible ways in which the receptor output could be coding the message:

(1) *Simple code*: According to this code, different messages would correspond to different average *frequencies*; whenever the average frequency in the two axons is the same, the message is the same (Fig. 9-14). If the system is sensitive to differences in frequency of 5 pulses per second, the following messages will be possible:

MESSAGES	PULSES/SEC
1	10
2	15
3	20
.	.
.	.
.	.
19	95
20	100

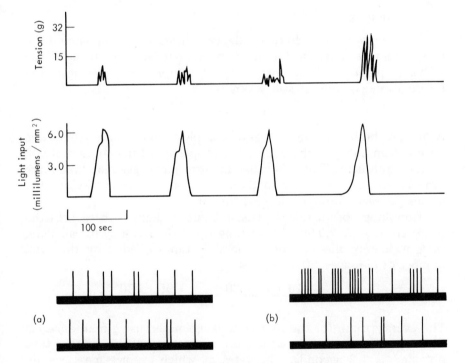

Fig. 9-14. The crayfish photoreceptor system studied by L. Stark and others detects light as an input and gives a tension at the tail as an output. Two possible ways in which the information might be coded are shown. In (a) there is a response when both axons fire at the same average frequency; in (b) the response occurs if both axons fire at the same time.

We assume that the 20 messages are equally probable (the probability of having any one message appearing is 1/20). The information content is given by:

$$H = -\frac{1}{20} \log_2 \frac{1}{20}$$

$$= -\frac{1}{20} \log_2 \frac{1}{20} - \frac{1}{20} \log_2 \frac{1}{20} - \ldots - \frac{1}{20} \log_2 \frac{1}{20} \, [20 \text{ terms}]$$

$$= -20 \times \frac{1}{20} \log_2 \frac{1}{20}$$

$$= -\log_2 \frac{1}{20}$$

$$= \log_2 20$$

(2) *Complex code*: In this code, two messages are the same if the pulses come exactly at the same time; that is, exact firing is the relevant quantity. Since there are 20 equiprobable messages, the information content is still given by

$$H = \log_2 20$$

Although both the simple and complex codes have the same information content, there are other properties of the codes that will not be equivalent. This is because an average message may take more time to transmit using one of the codes than the other; the number of bits per second, then, will be different.

How long does it take to transmit the message with each of the codes given above? The average time for code 2 is easy to calculate, as it is merely the sum of all possible times divided by the total number of messages, 20; this gives

$$\text{average time (2)} = \frac{10 + 15 + 20 + \ldots + 95 + 100}{20} = 50 \text{ msec}$$

The average time for the first code is estimated using mathematical techniques that are somewhat more sophisticated; the number thus obtained is 1.2 seconds. The rates at which average messages are transmitted is, then,

$$\frac{\log_2 20}{1.2} = 3 \text{ bits/sec}$$

for the first code, and

$$\frac{\log_2 20}{0.05} = 100 \ \text{bits/sec}$$

for the second code.

This quantity, given in bits per unit time, is called the *capacity of the channel*. If we had a way of measuring the capacity, we could decide between one code and the other. Fortunately, the crayfish provides a unique opportunity for studying this problem because it has essentially two separate channels that are completely equivalent from the functional point of view: It is possible to cut either axon and obtain the same output. If, for example, during a given second, one of the channels (axons) transmits a certain signal and the second channel transmits a signal which contains some, but not all, of the elements of the first, we can compare the code words that are the same in both channels and conclude that the message is given by the elements common to both channels, the rest being irrelevant signals (noise). This is achieved using a special mathematical trick called cross-correlation, which tells which elements two signals have in common. In practice, recordings are obtained from both axons and also from each individual axon; the results obtained by Stark show that only about 3 bits per second are relevant. The simple, average-frequency code is, then, the most likely candidate for carrying the photoreceptor information. We should point out that other possible 3-bit codes are all equivalent to the average-frequency code considered here. We conclude, then, that the frequency, not the precise pattern, of the firing is important and that about 97 bits are wasted in noise.

9-11. Is This Coding Efficient?

The biological question may arise as to why such an "inefficient" system was not de-selected; that is, why should so many bits be "wasted"? Actually, the situation is not wasteful. If 100 bits per second were used, the system itself would be required to have absolute knowledge as to how to recognize the 100 bits, and the genetic message would be required to *waste* information in this knowledge. Furthermore, notice that the difference in average transmission time for the two codes is not significant, so the "cheaper" code (the 3-bit code) is more advantageous from every point of view.

10

THERMODYNAMIC EFFICIENCY, BIOLOGICAL AND MECHANICAL MACHINES

. . . the animal frame, though designed to fulfill so many other ends is a machine more perfect than the best contrived steam engine—that is, capable of more work with the same expenditure of fuel.—James Prescott Joule (1818-1889), businessman, scientist, and beer brewer.

10-1. Efficiency of Heat Engines

Had Sadi Carnot lived in the 1970's, his book *"Reflections on the Motive Power of Fire"* might very well have been a collection of poems on the occult. Most likely, it would have been read as often as his theoretical proof that the maximum efficiency of a cyclic engine driven by heat can at most be (Prob. 10-1):

$$(10\text{-}1) \qquad E = \frac{T_1 - T_0}{T_1}$$

in which T_1 is the warmest temperature reached by the engine during a cycle, T_0 is the coldest temperature, and E is the efficiency of the engine (defined as the ratio of the work output to the heat input). Since negative temperatures are ruled out from the absolute scale used in Eq. 10-1, the efficiency of such an engine is less than one.

Carnot also showed that in order to achieve this efficiency the machine would have to follow the series of steps depicted in Fig. 10-1: Expansion at constant (warm) temperature, expansion without exchange of heat with the environment (adiabatic), compression at

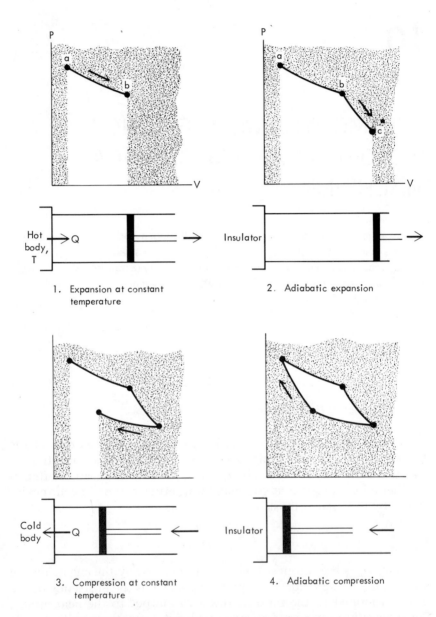

Fig. 10-1. A Carnot cycle comprises an expansion at high temperature, an adiabatic expansion, a low-temperature compression, and an adiabatic compression. It is the most efficient way of obtaining work from a heat engine, but there is no realizable practical machine that can approximate a Carnot engine.

constant (cold) temperature, and compression without exchange of heat (adiabatic). From Eq. 10-1 we can conclude that the higher the warm temperature and the lower the cold temperature, the more efficient the machine will be. This conclusion is correct, but a Carnot machine cannot be built. The main practical reason was discovered by the inventor of the Diesel engine (Rudolph Diesel), who found out that it is practically impossible to deliver heat at constant temperature. From a thermodynamic point of view, it is irrelevant whether the machine can be built or not: The important point is that as far as conversion of heat into mechanical work is concerned, no machine can exceed the efficiency of Carnot's ideal engine.

It is for this reason that Carnot stated the Second Law of Thermodynamics in terms of the impossibility of converting all the heat added to a (heat) engine into work.

10-2. Two Obvious Problems

There are two obvious problems in applying Carnot's considerations to the efficiency of real engines in general and to biological machines in particular. The *first* problem is that not all machines transform heat into work. For example, we have seen that living things utilize chemical, rather than thermal, energy. In fact, the only enlightening point Carnot's treatment provides is that a biological entity which transformed thermal energy into work directly would cook before it had a chance to show how efficient it was (Prob. 10-2).

The *second*, more serious, problem has to do with the static structure of classical thermodynamics. While real engines work at a finite rate, in order to analyze them we must pretend that they are moving infinitely slowly. In doing this we eliminate the variable "time" completely from the calculations of reversible or equilibrium thermodynamics. There is absolutely no reference to time except, perhaps, for the implicit dictum "time must have died by the time we perform a measurement." The machines of thermodynamics, then, either work infinitely slowly (in which case they are useless) or irreversibly (in which case we cannot calculate the work done).

Some man-made machines do work close to equilibrium, but this is the exception; an example is a lever (Fig. 10-2), which is used to lift a weight W. In this or in the similar case in which we use a hoist to remove the engine from a car, time considerations are irrelevant: The important problem is to lift the weight no matter how long it takes. The amount of work done on both sides of the pivot is the same and the process is therefore reversible.

Fig. 10-2. An equilibrium engine: a lever. This type of machine is used to lift heavy weights very slowly; time considerations are irrelevant, and the system is essentially in equilibrium.

10-3. The Way Around the Equilibrium Problem: Local Equilibrium and Steady State

If we look back to the diffusion treatment given in Chapter 9, it will become clear that at no time did we give a thermodynamic justification for the introduction of motion (flow of matter). The treatment is, however, valid because all potentials and, implicitly, all other intensive variables such as P and T were assumed to be constant in time and because the steady-state requirement assured that the extensive properties such as mass would also be constant in every region of the system. Thus, the system does not depart considerably from equilibrium.

A simple example is depicted in Fig. 10-3, in which a membrane separates two compartments kept at different hydrostatic pressures.

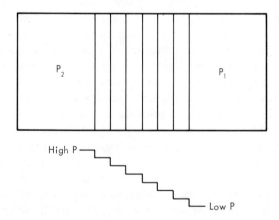

Fig. 10-3. When the pressure differences between two compartments separated by a membrane are small, it can be assumed that the change in pressure takes place in very small steps across the membrane so that a uniform pressure—without gradients—exists at each point, and quasi equilibrium prevails.

The difference in pressure is an external *force* that displaces the particles inside the membrane from their equilibrium state, thus leading to motion of these particles. Imagine that the membrane is subdivided into thin vertical strips of equal width. The total difference in pressure can be assumed to be uniformly distributed across these strips so that the sum of all the pressures add up to the overall pressure difference that exists between the two compartments. If we denote the overall pressure difference by $P_2 - P_1$ and the number of partitions by N, the pressure across each small region in which we have subdivided the membrane is

$$\Delta P = \frac{P_2 - P_1}{N} .$$

If N is a large number and $P_2 - P_1$ is small, ΔP can approach zero, the equilibrium state—the pressure in the subcompartment is everywhere (about) the same. Aside from the assumption that there are enough molecules in a region, we have to make sure that the number of molecules in each stays constant in time; this is only possible in the steady state, as one molecule enters a compartment whenever another molecule leaves it (Fig. 10-4). With these restrictions in mind, we can now write *time* changes in functions of state, as

$$\frac{\Delta S}{\Delta t} , \frac{\Delta E}{\Delta t} , \text{etc.}$$

The way to maintain the flows constant (in the steady state) is, then, to fix the values of the intensive variables in the system (concentration charge, etc.). When this is done, open systems approach the steady state; otherwise they need not.

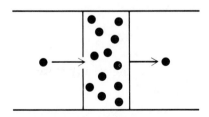

Fig. 10-4. In a situation of steady-state flow, molecules enter and leave a region at the same rate.

10-4. Energy vs. Power

In the dynamic (steady) state the work done *per unit time*, rather than the work done in a cycle, is of interest. The relevant quantity is now not energy but *power*, the time rate of change in energy.

An equivalent expression for steady-state power is easily found. Consider an object moving with constant velocity v when a driving force F acts on it.* From the definition of power,

$$(10\text{-}2) \qquad\qquad \text{Power} = \frac{\text{work}}{\text{time}}.$$

Furthermore, since work $= F \times d$ for a constant force F, we can rewrite this equation as

$$p = \frac{F \times d}{t}.$$

But the ratio d/t is simply the velocity v, so the final expression for p can be given in the form

$$(10\text{-}3) \qquad\qquad p = F \times v.$$

Clearly, if N objects move with the same velocity v under the action of equal forces F, the total power will be

$$(10\text{-}4) \qquad\qquad p = F \times v \times N.$$

For the special case of diffusional flows considered in Chapter 9, we could calculate the power used to move the molecules through some arbitrary volume of the membrane. Call this volume V_0. The power used is, then,

$$\frac{p}{V_0} = F \times v \times \frac{N_0}{V_0},$$

in which N_0 is the number of molecules in the same volume. Since the ratio N_0/V_0 is simply the concentration c carried by the diffusional flow, we can write

$$\frac{p}{V_0} = F \times v \times c.$$

But we saw in Sec. 9-7 that the steady-state diffusional flow is given by $v \times c$. It then follows that

---

*At the risk of repetition: The velocity v can only be constant if the *net* force is zero, so there must be a frictional force opposite to the direction of motion to balance F out.

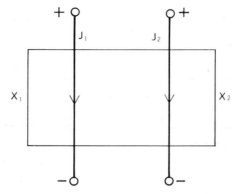

Fig. 10-5. An energy transducer may be described schematically by specifying the input and output forces and flows.

(10-5)
$$\frac{p}{V_0} = F \times J,$$

in which F is the driving force $F = -\Delta\mu/\Delta x$.

10-5. Transducers

How do we define a machine in a very general sense so that we can include more than heat or a specialized design? The idea of a *transducer* comes in handy: A transducer is an energy converter which transforms any kind of energy into any other kind of energy and which has an input and an output. It is usually represented by means of a "black box," as shown in Fig. 10-5.

There are two general types of transducers: A *bilateral* transducer transforms energy in either direction (input to output or output to input); a *unilateral* transducer transforms energy in only one possible direction.

The eye is a unilateral transducer which transforms electromagnetic energy (input) into a series of spikes whose frequency is related to the intensity, frequency, and other properties of the stimulus, depending on the complexity and degree of sophistication of the given eye. Obviously, the eye cannot be used in reverse to emit light.

An electric motor provides a simple example of a bilateral transducer. If a voltage difference is applied to the electrical terminals of the motor (input), its central axis rotates (output); conversely, electrical energy is generated when the axis of the motor is forced to rotate by mechanical means (Fig. 10-6).

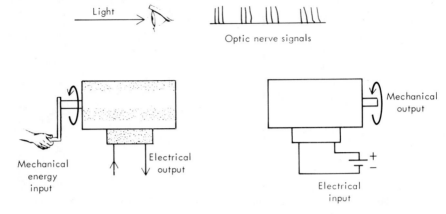

Fig. 10-6. The eye is a unilateral transducer that converts light into electrical signals. A motor is a bilateral transducer that may be used to convert mechanical to electrical energy.

The transducer concept is particularly useful in the analysis of biological systems because we very rarely have available a complete mechanistic description of the system and because given two experiments in which the same number and type of measurements are made and the same information is obtained, the best one, from the functional point of view, is the more peripheral one; that is, the one that disrupts the normal functioning of the biological system in a minimal way.

Some common biological transducers are given in Table 10-1.

TABLE 10-1. Typical Biological Transducers

TRANSFORMATION	ORGAN TRANSDUCER
Chemical to electrical energy	Brain, nerve, nose, tongue
Chemical to mechanical energy	Muscle
Chemical to osmotic energy	Kidney and all cell membranes
Chemical to radiant energy	Firefly luminescent organ
Light to chemical energy	Chloroplast
Light to electrical energy	Eye
Hydrostatic to electrical energy	Inner ear
Sound to electrical energy	Ear

Source: After T. P. Bennett and E. Frieden, *Modern Topics in Biochemistry* (New York: Macmillan Co., 1967).

10-6. Biological Processes Consist of Many Transducers Attached to One Another

Most biological energy transformations involve the conversion of chemical to chemical or chemical to mechanical energy; there is, however, a very sophisticated type of energy conversion that takes place in the nervous system (which changes electrical signals into meaningful responses) and in the biosynthetic system of the cell (which transforms the program of the gene into biologically meaningful structures), such as the examples considered in the information chapter.

Biological "engines" consist of several transducing steps; in these, the output of one process becomes the input of the next process. Take, for example, the emission of light by a firefly. Some of the steps involved are indicated in Fig. 10-7. It can be seen that three different kinds of energy conversion are involved: there is first an interaction between electrical and chemical energy (release of acetylcholine from nerve), followed by a series of chemical steps which lead to the oxidation of a substance called luciferin, with the emission of light.

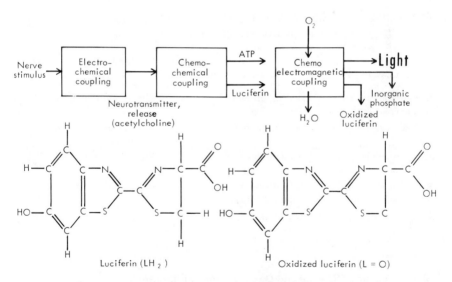

Fig. 10-7. Main steps that lead to the emission of light by the firefly. (For a detailed account, see W. D. McElroy and H. H. Seliger, "Biological Luminescence," *Scientific American*, December 1962.)

10-7. How Can Transducers Be Described in Quantitative Terms?

In dynamic transducers, the quantities considered are input power and output power; if the system is in the steady state, it is not necessary to give the heat dissipated, as this can be found from the Second Law. Although the simplest possible way of specifying the system behavior would be to give the drawing:

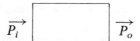

in which P_i and P_o are the input and output powers, respectively, a more complete description involves giving input and output powers in terms of forces and flows. The main problem is to find the explicit expression for power in terms of a force-flow product for each form of energy. As before, we must start with the expression for work. Typical examples are calculated in Prob. 10-3 and summarized in Table 10-2. The force, X, which gives rise to a flow, J, is called the force conjugate to that flow.

TABLE 10-2. Some Important Conjugate Flows and Forces
Considered by Nonequilibrium Thermodynamics

FLOW J	CONJUGATE FORCE X	NAME OF PROCESS	POWER TERM
1. Reaction rate per unit volume, J_{chem}	Chemical affinity, A	Chemical reaction	$A \times J_{chem}$
2. Electrical current, I	Electrical potential, E, $\dfrac{\Delta \Psi}{\Delta x}$	Flow of electric current	$I \times E$
3. J_i flow of uncharged species i	Chemical potential gradient, $-\Delta\mu/\Delta x$	Flow of matter	$J_i \left(\dfrac{\Delta\mu}{\Delta x}\right)$
4. J_i flow of charged species i	Electrochemical potential gradient, $-\Delta\tilde{\mu}/\Delta x$	Ionic flow	$J_i \left(\dfrac{\Delta\tilde{\mu}}{\Delta x}\right)$
5. J_V volume flow	Hydrostatic pressure difference, ΔP	Volume or bulk flow	$J_V \Delta P$

10-8. Dynamic Efficiency and Onsager (Nonequilibrium) Thermodynamics

In the early 1930's physical chemist Lärs Onsager gave a thermodynamic formulation that attempted to describe the behavior of interacting steady-state processes. Although the generality of this approach has been questioned many times, it is worth taking into consideration because it is extensively used in the physiological literature.

Basically, Onsager related the time change of internal entropy production in a given transducer to the input and output powers (see Prob. 10-4),

$$(10\text{-}6) \qquad T\frac{\Delta_i S}{\Delta t} = \text{power in} - \text{power out}$$

$$= X_1 J_1 - X_2 J_2,*$$

in which J_1 and J_2 are the input and output flows, and X_1 and X_2 the input and output forces. The expression $T\,(\Delta_i S/\Delta t)$ is called the *dissipation function* and is denoted by Φ.

Onsager was able to show that in many cases the input and output flows are related to the forces by linear equations of the form

$$(10\text{-}7) \qquad J_1 = L_{11} X_1 + L_{12} X_2$$

$$J_2 = L_{21} X_1 + L_{22} X_2$$

in which L_{11}, L_{12}, L_{21}, and L_{22} are numbers characteristic of the given transducer. These *phenomenological equations* indicate that the input and output forces are a function of *both* their conjugate *and* nonconjugate flows. J_1, for example, is a function of both X_1 and X_2 rather than X_1 alone. Furthermore, Onsager proved that the contribution of X_2 to the nonconjugate flow J_1 is equal to the contribution of X_1 to the nonconjugate flow J_2, so that $L_{12} = L_{21}$. We have already given an example of such a set of equations in Eq. 7.40,

$$J_V = L_P\,\Delta P - \sigma L_P RT\,\Delta c$$

$$J_D = -\sigma L_P\,\Delta P + L_D RT\,\Delta c.$$

In these equations, the forces and flows are

$$J_1 = J_V,\, J_2 = J_D$$

$$X_1 = \Delta P,\, X_2 = RT\,\Delta c$$

---

*In most of the nonequilibrium thermodynamics literature, a plus—rather than minus—sign appears in this expression. That indicates that all the power contributions are considered as inputs and no distinction is made between input and output.

while the phenomenological coefficients are

$$L_{11} = L_P, L_{22} = L_D$$

and

$$L_{12} = L_{21} = -\sigma L_P.$$

10-9. Spontaneous Coupling of Two Processes Requires Positive Entropy Production

Since entropy increases in time in a spontaneous irreversible process, it follows that the expression $\Delta_i S/\Delta t$ must be positive; that is,

$$\frac{\Delta_i S}{\Delta t} > 0.$$

It then follows that the dissipation function is also positive for an irreversible process,

$$\Phi > 0.$$

For a *reversible* process, entropy remains constant and $\Phi = 0$.

The requirement of positive entropy production is, of course, equivalent to the statement that the input power must be larger than the output power. In the reversible case, input and output powers are equal.

10-10. How Nonequilibrium Thermodynamics Changes Our Perception of Some Biological Problems

Unlike classical thermodynamics, Onsager's approach stresses the importance of the *rates* at which processes take place as well as the generality of interactions *between* processes. As a result, many biophysical and biochemical steps need reinterpretation in the light of this theory.

Consider, for example, the active transport of n moles of an uncharged substance across a chemical potential difference $\Delta\mu$. According to thermostatics, the metabolic step (splitting of ATP, for example) needs to provide a free-energy decrease

$$\Delta G = n \, \Delta\mu$$

or larger. Nonequilibrium thermodynamics, on the other hand, says that this expression is meaningless unless we include the flow rates of the substance that goes through the membrane and the *rate* of the metabolic reaction. Problems at the end of this chapter will consider specific numerical examples.

10-11. Coupling and Asymmetry

Not all kinds of forces and flows can interact to form transducers. If the input and output processes are of the same kind, they will in general interact. But how about if they are different? Can a chemical reaction induce the transport of matter, for example? We know that this type of interaction exists in the form of active transport. Since active transport depends on the hydrolysis of ATP, there must be a transducing step which relates the rate of this chemical step to the motion of matter across a membrane.

While the molecular basis of active transport is not understood, we know that coupling between the chemical and transport steps is not possible unless there exist spatial asymmetries in the given membrane. The reason can be easily understood by considering two possible ways of building a rocket. In the first, a chemical propellant is placed in the middle of a tube open at both ends. Clearly, the forces on the top will cancel out the forces on the bottom, and the "rocket" will not move. In the second option, one of the ends of the tube is closed; when the propellant goes off the rocket will move. The possibility of coupling the chemical reaction to the motion arose by introduction of a physical asymmetry. Similarly, coupling between transport of matter and metabolism will not take place in a biological membrane unless there is an asymmetry built into the membrane.

The late Aaron Katchalsky of the Weizmann Institute was fond of demonstrating a machine invented by A. Oplatka which illustrates how asymmetry can lead to observable chemochemical coupling. In essence, the machine is based on the observation that collagen fibers contract in the presence of a concentrated salt solution and expand when washed with water. The engine has a collagen belt that goes from a water compartment to a concentrated salt (LiBr) solution; thus, the fibers contract in one compartment and relax in the other. If the system were perfectly symmetrical, the machine would never move because all the forces would cancel out; the introduction of an inelastic coupling belt, however, introduces a torque that keeps the engine running (Fig. 10-8).

This is a striking demonstration both because it actually shows the coupling and because it looks as though the machine could run

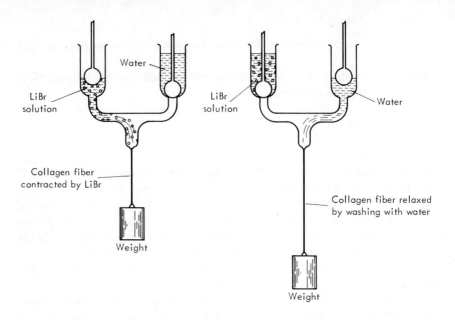

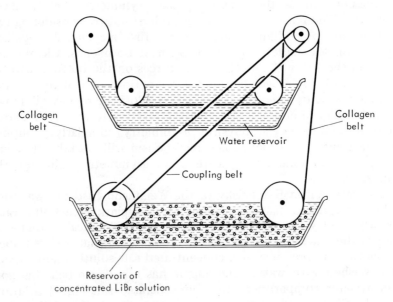

Fig. 10-8. Laboratory demonstration of chemomechanical coupling. A collagen fiber contracts in the presence of a concentrated salt (LiBr) solution and relaxes by washing in water. A cyclic machine in which collagen circulates between a salt bath and a water reservoir can be made by providing an asymmetrical coupling belt. (A. Katchalsky, *Chemical Dynamics of Macromolecules and Its Cybernetic Significance in Biology and the Physical Sciences*, ed. by S. Devons. New York: Columbia University Press, 1969.)

forever utilizing the concentration difference. Actually, the collagen fibers carry salt from the concentrated reservoir to the pure water solution and effectively "discharge" the system: when both concentrations are equal, all motion ceases.

Problems

10-1. Show that the Carnot engine has the efficiency

$$E = \frac{T_1 - T_0}{T_1},$$

in which T_1 is the high temperature and T_0 the low temperature.

Hint: Follow these steps:

(*a*) Consider the change in entropy for each of the four parts of the cycle: expansion at constant temperature T_1, adiabatic expansion, compression at T_0, adiabatic compression.

(*b*) For each of the steps in (*a*) calculate the heat gained or lost, and express these in terms of temperatures and entropy changes.

(*c*) Express the efficiency as the output work divided by the heat taken from the high temperature reservoir and assume that the process is reversible.

10-2. Utilize the result obtained above to find the external temperature at which the human body ($37°$ C) would have to work if it were a thermal engine with a conversion efficiency of 20%. Discuss.

10-3. Starting from the definition of power for a constant force, $p = (f \times d)/t$, give expressions for

(*a*) The power required to transport mass at a rate J between two compartments separated by a distance d,

(*b*) The power required to displace a volume of gas at a rate J_v applying a pressure P,

(*c*) The power required to move a charge at a rate I between two regions at different electrical potential.

10-4. Relate the dissipation function to the time decrease of free energy.

10-5. What is the rate of internal entropy production when a substance moves at the rate of 0.1 mole/sec from a compartment at which it is at a concentration 2 moles/liter to a compartment at concentration I assuming ideal conditions and a temperature of $27°$ C?

10-6. What is the increase in entropy during 3 seconds in the system considered in the previous example?

10-7. A two-compartment system has the following measurable properties:

Right compartment:	glucose = 1 molar; sucrose = 0.3 molar
Left compartment	glucose = 0.2 molar; sucrose = 0.1 molar
Left to right flow:	glucose = 0.025 mole/sec cm^2
	sucrose = 0.01 mole/sec cm^2

Calculate the dissipation function, Φ, for this system. Is this a passive process?

10-8. A solution of 0.1 molar sucrose is filtered through a membrane having an area of 1 cm^2. The filtrate is found to have a sucrose concentration of 0.025 mole. The hydraulic conductivity of the membrane is found to be 0.79×10^{-6} cm^3/dyne · sec. What rate of volume flow will be observed when the membrane separates a 0.1 molar sucrose solution from a compartment filled with distilled H_2O?

11

THERMODYNAMIC EFFICIENCY
AND ECOLOGY

Ecology is the part of biology which deals with an organism in relation to its environment. As the essence of this interaction is the continuous exchange of mass and energy that takes place between the biotic and abiotic worlds as well as within the biosphere, thermodynamic laws play an important role in ecological considerations. It makes little sense, for example, to talk about osmoregulatory mechanisms (see Sec. 6-20) without considering the osmotic environment in which the given organism lives. In the following discussion, we restrict our attention to a few basic ecological examples with strong thermodynamic connotations: food chains, pollution, and ecological successions.

11-1. Efficiency of Energy Flow in the Biosphere: Implications from the Second Law of Thermodynamics

We indicated in Chapter 1 that the functioning of the biosphere depends on the continuous trapping of the energy arriving from the sun. Once green plants achieve this step, energy is transferred to herbivores and from herbivores to carnivores. This series of mass and energy transfers receives the name of *food chain*.

Some food chains have been extensively analyzed and can now be described in quantitative terms; a convenient description is provided by an energy-flow diagram such as the one given in Fig. 11-1 for the ecosystem at Silver Springs, Florida, studied by ecologist Howard T. Odum.

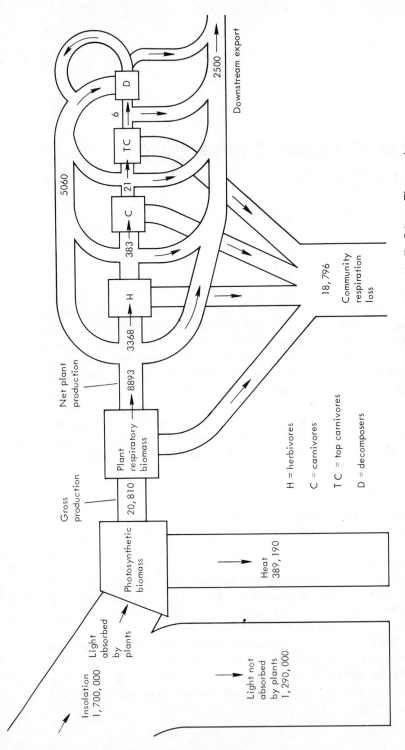

Fig. 11-1. The ecosystem of Silver Springs, Florida, described by H. T. Odum. The numbers indicate the amount of energy (in kilogram calories per square meter per year) that flows between different levels of production and consumption. (Based on drawing by H. T. Odum, *Ecological Monographs*, 1957, 27, 55.)

At each box (transducer) the input energy, energy lost in the form of heat, and energy transferred to the next trophic level are indicated. Two simple results predicted by thermodynamics are obvious from this flow graph.

First, for a system in the *steady state*, the energy that enters a given box equals the total energy that leaves it. Thus, the total energy intake by herbivores in one year, 3,368 kilogram calories per square meter, equals the energy "produced" by the herbivores—the energy acquired by the primary carnivores, 1,478 kilogram calories—plus the energy lost in the form of heat during respiration, 1,890 kilogram calories.

Second, the amount of energy transferred to a given trophic level is smaller than the energy received by the previous trophic level. For example, the total intake by plants is 410,000 kilogram calories; the energy input to herbivores is 3,368 kilogram calories; and top carnivore intake is only 21 kilogram calories per square meter a year. This inefficiency in handling energy is, of course, predicted by the Second Law of Thermodynamics, and it is most strikingly demonstrated in the form of a "pyramid of energy" such as is shown in Fig. 11-2. The pyramid of energy is obtained by assigning bars whose lengths represent the amount of energy received by each trophic level per unit time. While the pyramid of energy is not as complete as the flow diagram (it is not possible to indicate the energy intake by primary producers, for example), it provides a pictorial representation of the Second Law in action. It should become clear from the drawing why top carnivores are a rare luxury: Large amounts of primary production—hence, large numbers of plants—are needed to maintain a single top carnivore. This is efficiently conveyed by the "pyramid of numbers" in which the numbers of individuals at each trophic level is indicated by a bar

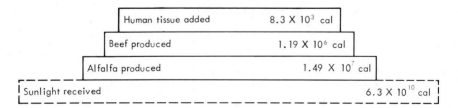

Fig. 11-2. Example showing how the pyramid of energy arises as energy is transferred from producers to consumers; the amount available for utilization decreases as predicted by the Second Law of Thermodynamics. (E. P. Odum, *Fundamentals of Ecology*. Philadelphia: W. B. Saunders Co., 1953.)

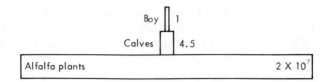

Fig. 11-3. Pyramid of numbers for a simple food chain. (E. P. Odum, *Fundamentals of Ecology*. Philadelphia: W. B. Saunders Co., 1953.)

whose length is proportional to the population at that level (Fig. 11-3). The pyramid of numbers is, however, misleading, because it incorporates not only the effect predicted by the entropy principle but also the relative metabolic rates for organisms of different sizes. Metabolic rates of mammals (per unit weight), for example, decrease with increasing size (Fig. 11-4); small mammals must eat large

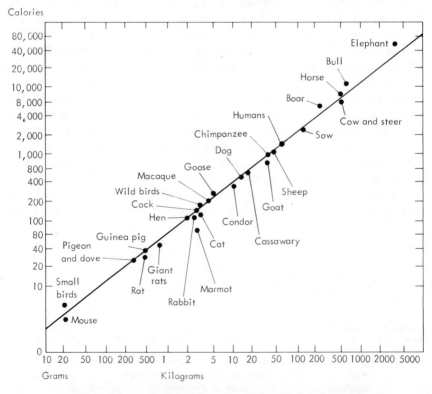

Fig. 11-4. Relation between heat production and body weight in mammals. (D. W. Bishop in *Comparative Animal Physiology*, ed. by C. L. Prosser. Philadelphia: W. B. Saunders, 1950.)

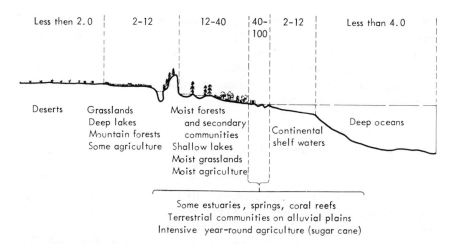

Fig. 11-5. World distribution of primary energy production. (E. J. Kormondy, *Concepts of Ecology*. Englewood Cliffs, N.J.: Prentice-Hall, Inc., 1969. Adapted from E. P. Odum, *Fundamentals of Ecology*.)

amounts of food very fast. Physiologist Knut Schmidt-Nielsen points out that the smallest mammal, the shrew, has such a high metabolic rate that it must eat almost the equivalent of its own body weight in a day. The *total* metabolic rates are, however, larger for larger animals; the energy input required by a large animal is larger than that required by two animals each half the size of the large one.

11-2. What Are the Relative Efficiencies of Producers and Consumers?

There are no general rules from which the efficiency of energy conversion in a given step of a food chain can be found. Table 11-1 gives three examples for aquatic ecosystems, including Silver Springs.

TABLE 11-1. Efficiency of Energy Transfer in Various Ecosystems

TROPHIC LEVEL	CEDAR BOG LAKE, MINN.	LAKE MENDOTA, WIS.	SILVER SPRINGS, FLA.
Producers	0.10	0.40	1.2
Herbivores	13.30	8.70	16.0
Small Carnivores	22.30	5.50	11.0
Large Carnivores	–	13.00	6.0

Source: E. P. Odum, *Fundamentals of Ecology* (Philadelphia: W. B. Saunders, 1953).

The only generalization that seems to emerge from this table is that the primary step in energy "production" is the most inefficient. Some of the possible reasons for this tendency will be discussed when we compare efficiency with output power. The distribution of gross primary energy productions are indicated, for the sake of completeness, in Fig. 11-5.

11-3. Pollution and the Smogmobile

Cartoonist Al Capp once introduced in his strip "Little Abner" an interesting invention which looked rather improbable even to those least informed on the detailed aspects of automobile technology. The device under consideration was called a smogmobile and it had the unlikely property of running on smog, as shown in Fig. 11-6. The reason why it is not possible to build such a machine is deeply rooted in the Second Law of Thermodynamics and has nothing to do with common sense: If smog were not a degraded form of energy—if it

Fig. 11-6. Al Capp's Smogmobile. (*Boston Sunday Globe*, July 12, 1970; New York News, Inc. All rights reserved.

were composed of gaseous fuel, for example—there would be nothing unreasonable about using it to drive a car or any other engine. Furthermore, if the arrow in the drawing were reversed, the process would look perfectly normal; the original problem, then, lies in attempting to run an irreversible process in the wrong direction without providing an external source of energy. Whatever Al Capp's reason for introducing such a device, the smogmobile points to the central thermodynamic problem of industrial pollution: the limitations imposed by the entropy principle.

11-4. Pollution and Thermodynamics

The basic problem of environmental pollution is due to thermodynamic restrictions imposed by nature. Pollution is a direct consequence of the inefficient use of energy because of

- The utilization of degraded energy to generate more "pure" forms of energy (this is related to thermodynamic efficiency)
- Requirement of large quantities of power by the general public for enjoyment of the comforts of modern society.

To these basic restrictions, which result from combining the limitations of physical laws with the energy requirements of our civilization, we must add:

- The utilization of poorly designed converters (an engineering problem)
- The unwillingness of producers to salvage waste products (an economic-thermodynamic problem)

The last two restrictions are not absolute, and can be improved upon with the tools provided by modern engineering.

11-5. Second Law and Pollution

The Second Law tells us two things: First, the output power of any machine is less than the input power; second, the difference is lost in a form of energy which cannot be recovered without performing work. This is the part that "pollutes." In open systems there is both a *degradation of energy* to less available forms and a *degradation of mass*, so in general pollution consists of both mass and energy (Fig. 11-7).

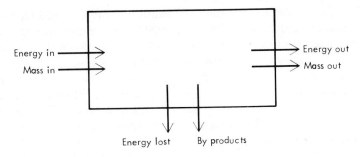

Fig. 11-7. Conversion of chemical energy involves the degradation of both energy and mass.

Recovery of wasted energy takes, according to the Second Law, more work than the amount originally lost (in an irreversible process). This means that recovery will in general be a costly process, although there are some interesting schemes for reusing waste products *without* bringing them to the original high-energy level.

All engines are inefficient, and the problem is present in any energy transformation. In the animal body, for example, energy conversion with subsequent excretion of waste products is a good example. But the dangerous levels are reached when large amounts of energy are inefficiently utilized. For this reason, industrial countries are the ones more seriously hit with this disease. It is interesting to point out that the energy processes that take place in human bodies change the environment by an infinitesimal amount, as opposed to plants—whose metabolic activities affect the environment directly. The reason civilizations can affect the environment to such an extent is because industrial societies do not handle energy directly; they only handle information. We have already pointed out that information is easier to amplify than energy. Man puts pieces of steel together in a way in which a program or memory is built in the structure. That is basically what a motor or any other energy-converting structure is.

11-6. On Technical, Conversion, and Application Efficiencies

In principle, the limitations of energy-conversion efficiency are dictated by whether the system changes a high-quality energy into a degraded form or vice versa and on the relative quality of the two energies. As we pointed out, gravitational energy has very pure quality, and its conversion into any other forms of energy can be effected efficiently. Since mechanical energy can also be converted to electrical energy efficiently, two transducers that transform gravitational to mechanical to electrical energy work with an efficiency close to 100%. This is what happens at a hydroelectric plant in which the gravitational energy of a waterfall is converted into the mechanical motion of rotation of a motor which generates electricity. Similarly, electrical energy can be converted into heat with an efficiency of almost 100%, because heat is a degraded form when compared to the high-quality electrical input.

In practice, there are limitations imposed by the engineering of the given engine, so a similar type of energy conversion may give different efficiencies, as shown in Fig. 11-8. Thermodynamic considerations only refer to the *maximum* efficiency that can be achieved by a given transducer.

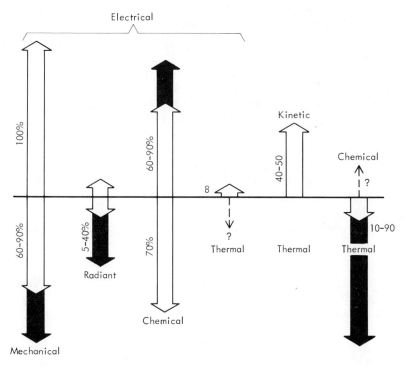

Fig. 11-8. Efficiency of energy conversion for various kinds of transducers. The arrows indicate the direction in which transduction takes place.

11-7. Gas vs. Electric Heat

When we attempt to reverse a transducer which normally transforms high-quality into low-quality energy, the machine becomes inefficient. For example, conversion of heat to electrical energy has an efficiency of only 8%—as opposed to 100% in the opposite direction!

With this general idea in mind, we consider the example given by Claude M. Summers in an interesting article in *Scientific American.* This is a comparison of the wasted energy that arises from converting natural gas into heat with the wasted energy that results from converting electrical energy into heat. If the electrical energy is obtained from gravitational power—waterfalls or tides—the efficiency is, as we said before, almost 100%. Unfortunately, not all electrical power is obtained this way. Rather, it is generated in "factories," where heat engines rotate electric generators and give electricity as an output. This part of the process is 32%. Although the application efficiency is large in the house heater (100%), the overall process is inefficient (32%). Figure 11-9 compares this with an alternative way of heating the house directly using natural gas (chemical energy) rather than generating electricity first.

This form of pollution, thermal pollution, *cannot* be completely cleaned. Even if the electrical factories were able to generate "smokeless" heat, the 68% of energy that is lost would still appear in the form of heat. It would then follow that the sole responsibility for polluting at the energy-providing steps rests on the shoulders of industrialists, but this is not the full story.

11-8. Power vs. Efficiency

We have so far assumed that the efficiency of an engine is constant. This could not be further from the truth: *The efficiency of a machine changes with the rate at which power is removed from the output.* Thus, a car going at high speeds is inefficient because it spends a lot of gas per mile traveled; its power output at the highest speed practically attainable is, however, maximal. At the other extreme, when the car is going very slowly, the efficiency is also poor (with the added disadvantage that the car does not go fast). The general curve for efficiency vs. power (which can be equated with speed in the case of the car) may look as shown in Fig. 11-10. It can be seen that at both the maximum and the minimum power outputs, efficiency is low and that there is some intermediate power value for which the efficiency is maximum.

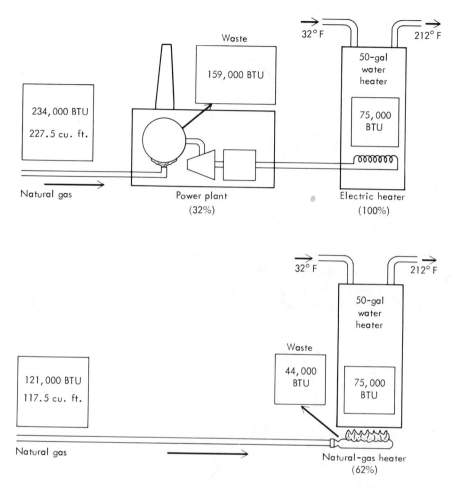

Fig. 11-9. Relative efficiencies for obtaining house heat through gas and electricity. (Adapted from the article by E. Cook, "The Flow of Energy in an Industrial Society," *Scientific American*, W. J. Freeman and Co., San Francisco, September 1971.)

A similar situation is seen in biological engines (Fig. 11-11). Eugene Odum gives an example for the case of algal cultures (*Chlorella, Scenedesmus*). High efficiencies are possible if the light intensity is low (10%, 20%, 50%), but the amount of food produced per unit time is low. If full daylight and large tanks are used, the efficiency drops to about 2% even though the production is maximal.

This result has an important practical implication; it indicates that a power supply company, for example, will work more or less

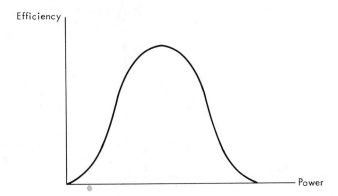

Fig. 11-10. The efficiency of a machine is a function of the rate at which energy is drawn from its output—it is a function of power.

efficiently depending on the rate at which the consumers demand power. The efficiency of the power generating unit, then, will depend on how many people want to watch T.V., use dishwashers, or listen to their hi-fi sets at a given time. We are all familiar with the

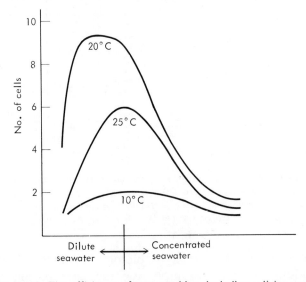

Fig. 11-11. The efficiency of any machine, including a living organism, depends on several external variables. As an example, the number of initial cell divisions of the sea plant *Porphyra tenera* can be seen to depend on both temperature and salt concentration. For a given temperature there is an optimum salinity that maximizes the rate of cell division.

summer power shortages when everybody turns on air conditioners (see Prob. 11-1).

One way to get around this problem is by building self-adapting machines that supply power at different rates depending on the number of consumers requiring power at a given time. This type of machine is not available in the present state of the engineering art. In biological systems, however, such self-adapting systems may exist. Thermodynamicist-biophysicist Roy Caplan, for example, has suggested that the muscle is such an engine. Although there is not enough experimental evidence at present to substantiate the claim, it would make sense that a perfectly designed muscle, which has to deliver various amounts of power depending on the load conditions, should have such self-adapting characteristics.

11-9. Can Pollution Be Completely Avoided?

It cannot. The Second Law of Thermodynamics clearly specifies that there is an inherent inefficiency in any machine we use. The idea, however, is to use the most efficient sources of power. Thus, one would want to transform gravitational energy (from waterfalls, tides, etc.) into other forms. The graph on the flow of energy in the United States in 1970 shows, however, that this energy represents only about 8% of the total work output for all kinds of energy (Fig. 11-12). The figure also shows that the energy lost in the generation and transmission of electricity is about 30% of the total energy waste even though electricity contributes only 10% of the total energy input.

Although energy waste is a price that civilization pays for the use of energy, this does not imply that the grand-scale pollution of streams and air is justified. We know that matter will degrade to less and less usable forms in the same way that energy does; unfortunately, some of the compounds happen to be very noxious to living things. While the Second Law states that this is unavoidable, it does not say that the products cannot be recovered or transformed into other—less dangerous or even useful—forms *provided that work is done*; that is, provided one is willing to pay for running machines that remove pollutants or special filtering systems that remove the pollutants but decrease the efficiency of the machines. Another approach to eliminating pollutants is, instead of destroying them, to utilize them for other industrial steps. Water is one of the many species that could be perfectly recovered through waste salvage.

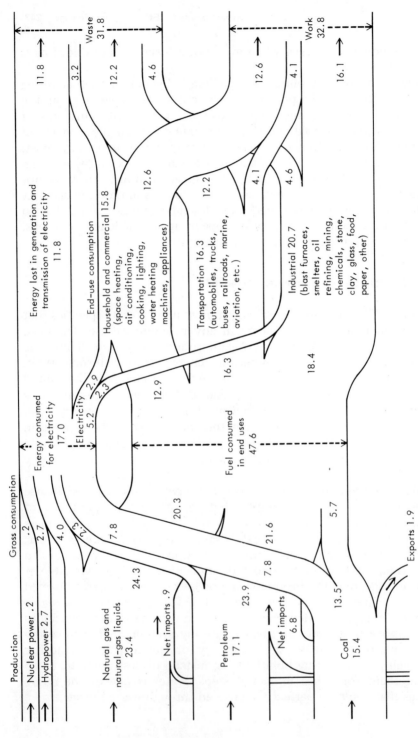

Fig. 11-12. Flow of energy in the United States during 1970. (Redrawn from E. Cook, "The Flow of Energy in an Industrial Society," *Scientific American*, W. J. Freeman and Co., San Francisco, September 1971.)

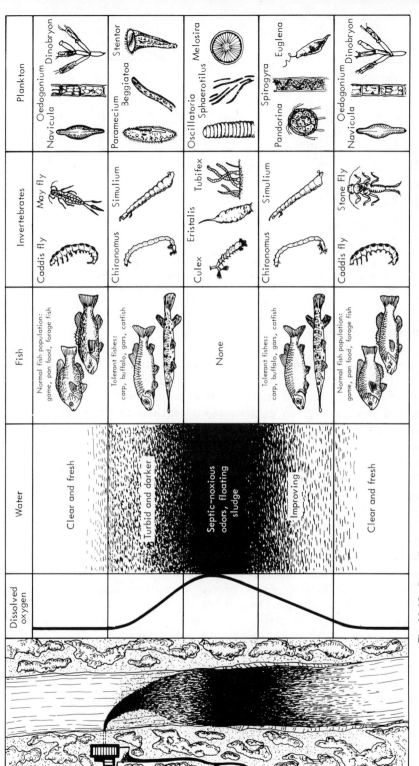

Fig. 11-13. Degrading matter has, in some instances, enough energy to do work of organization, as shown in this example in which distribution of organisms is given a function of their position relative to sewage plant. (R. Eliasen, "Stream Pollution," *Scientific American*, W. J. Freeman and Co., San Francisco, March 1952.)

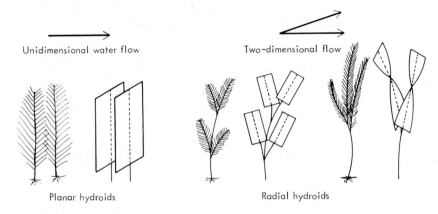

Fig. 11-14. Organisms adapt to match their energy needs with the characteristics of the environment. Hydroids, which obtain food by filtering water, are either planar or radial, depending on the geometry of the flow. (Adapted from Otto Kline, *Marine Ecology*, Wiley-Interscience, 1972.)

11-10. How Low Can You Get?

Degraded energy and mass can still produce the work of organizing information. An excellent example is provided in Fig. 11-13 in which the distribution of various organisms that live in the neighborhood of untreated sewage is given as a function of their distance from the polluting source. Clearly, the waste energy has been able to organize information. If there were no sewage, the biotic community would look everywhere the same. A direct degradation of energy—from motion to information—is seen in a similar situation in which the shape of hydroids adapts to the geometry of a water flow (Fig. 11-14). Hydroids extract food from water by filtration and the adopted shape also maximizes the rate of food acquisition. Thus, degraded energy can be used to *control* larger amounts of energy, in the same way that a light bulb can be turned on or off by a switch without spending large amounts of energy.

The most striking example of the organizing effect low-grade energy can have is found in the genetic mutation known as *industrial melanism*. H. B. D. Kettlewell showed that the spread of a dark (melanic) form of the peppered moth could be directly associated with the higher chances for survival of this mutant in industrial areas in which, because of pollution, tree barks became dark. The normal, white moth could easily be detected on the dark trees by predators. When pollution legislation was passed and most tree barks reverted to their light color, the normal white moth again became the predominant form (Fig. 11-15).

Fig. 11-15. The different colors of normal and melanic moths have been attributed to their differential concealment properties in nonindustrial and industrial areas. (Bruce Wallace and Adrian M. Srb, *Adaptation*, 2nd ed., © 1964. Reprinted by permission of Prentice-Hall, Inc., Englewood Cliffs, N.J.)

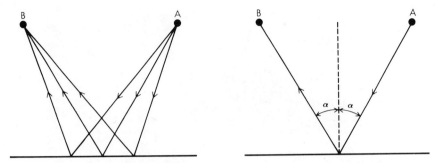

Fig. 11-16. A light ray going from A to B could reflect on a mirror following a number of possible paths. However, it only follows the path that requires less work (the angle of incidence equaling the angle of reflection). This is one of the many examples in which an observable phenomenon minimizes or maximizes a physical quantity such as energy.

11-11. The "Principle" of Minimum Entropy Production

We have seen that entropy increases in an isolated system until it reaches a maximum at equilibrium. Free energy, on the other hand, has its smallest possible value at equilibrium; in this case, we say free energy is a minimum. Quantities that can be maximized or minimized play an important role in physics. The power of the approach was first demonstrated by the "amateur" mathematician Pierre Fermat, who showed that the bending of light rays seen when light passes from a medium in which it travels at a speed v_1 to another in which it travels with a different speed v_2 minimizes the traveling time between two points in the two media (Fig. 11.16). In steady-state processes it appears that the quantity minimized is the dissipation function. Least dissipation (minimum free energy per unit time) seems to be a controlling factor in living systems.

At present, it is not known whether this principle is valid in general, but it does seem to be a basic law of the biological world. Most steady-state processes seem to have self-adjusted to achieve a state of maximum efficiency and minimum entropy production.

An example is provided by facultative respiration. Facultative organisms can generate ATP either aerobically (utilizing oxygen) or anaerobically (in the absence of oxygen). However, if they are placed in an oxygen atmosphere in which they have the "choice" of utilizing either step, they undergo the aerobic, more efficient, transformation.

In ecology, the phenomenon of succession is well documented: As time goes on, as the ecosystem becomes more mature, higher species tend to replace lower species, as exemplified in Fig. 11-17.

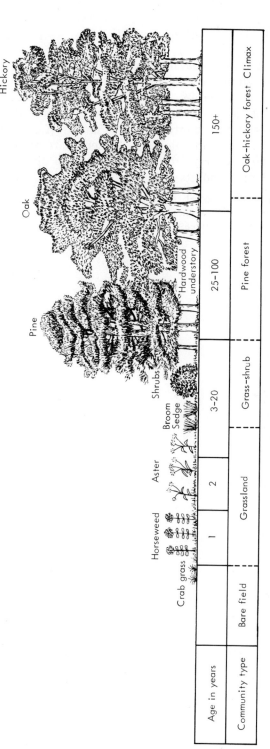

Fig. 11-17. Schematic example of succession given by ecologist E. P. Odum for the piedmont region of the southeastern United States. The transition to more mature forms reduces the productivity of the system but increases the efficiency of energy utilization. (E. P. Odum and H. T. Odum, *Fundamentals of Ecology*. Philadelphia: W. B. Saunders, 1950.)

With the passage of time, ecosystems also show longer food chains and closer relationships among the various members of the ecosystem. Although the mature system has a higher complexity, it maintains a constant flow of energy and a constant size of various populations. In this state, energy is much more efficiently used, because organisms in the food chain are "matched" to each other so that consumers acquire energy from producers at an optimum rate. Ecologist Ramón Margalef points out that

> An ecosystem has chances of survival with different degrees of organization, that is, with higher or with lower maturity. But the general trend is towards an increase in maturity.

Appendix

MATHEMATICAL CONCEPTS IN THERMOSTATICS AND NONEQUILIBRIUM THERMODYNAMICS

The Appendix introduces the calculus approach to thermodynamics and some of the more advanced applications and examples. To avoid repetition, conceptual ideas covered in the previous chapters are omitted. The Appendix should, then, be used as a supplement to the rest of the book rather than as an independent entity.

A-1. Reversible Work

Reversible work is done by following states that are infinitesimally close to equilibrium states. Refer to Fig. A-1, in which we consider a gas inside a piston. At equilibrium, the pressure exerted by the gas on the piston, P, equals the pressure exerted by the piston on the gas. As a result, there is no motion; it is possible, however, to increase the

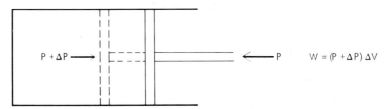

Fig. A-1. Infinitesimal work is calculated by assuming a small pressure increment and a small resultant displacement.

pressure of the gas by a small amount, ΔP, so that the piston will move. The work done by the gas is

$$W = (P + \Delta P) \, \Delta V.$$

If the states follwed in the transition are in the neighborhood of equilibrium states, the increment ΔP is infinitesimally small, so the (reversible) work is given by

$$W = P \, \Delta V.$$

Notice that since thermodynamics deals with transitions between equilibrium states, reversible work is the only type of work we can calculate.

A-2. Exact Differentials and Functions of State

Let z be a single-valued function of two variables x and y; that is, $z = z(x, y)$. For small changes in x and y, we can give the change in z in terms of the linear approximation

(A-1) $$\Delta z = M \, \Delta x + N \, \Delta y.$$

To find out what M and N are, we set $\Delta y = 0$ (that is, y is held constant), so that Δz becomes

$$\Delta z_{(y \text{ constant})} = M \, \Delta x$$

or
$$\left(\frac{\Delta z}{\Delta x} \right)_{(y \text{ constant})} = M.$$

When Δx goes to zero, M has the form of a derivative. Since we are taking the derivative of z with respect to x only, while y is held constant, we denote the "partial" character of this derivative by using a special symbol. Thus, we write the partial derivative of z with respect to x at constant y as

$$\left(\frac{\partial z}{\partial x} \right)_y = M.$$

Similarly, we can write the partial derivative of z with respect to y as

$$\left(\frac{\partial z}{\partial y} \right)_x = N$$

and in this case x, rather than y, is constant. When Δx and Δy become very small—when they approach the differentials dx, dy—Δz goes to the differential dz; that is,

(A-2)
$$dz = M\,dx + N\,dy$$

or, in terms of partial derivatives,

(A-3)
$$dz = \left(\frac{\partial z}{\partial x}\right)_y dx + \left(\frac{\partial z}{\partial y}\right)_x dy$$

Calculus shows (see Prob. A-1 as an example) that if z *is uniquely defined by x and y*, the partial derivative of M with respect to y equals the partial derivative of N with respect to x:

(A-4)
$$\left(\frac{\partial M}{\partial y}\right)_x = \left(\frac{\partial N}{\partial x}\right)_y.$$

In this case, we call dz an *exact differential*.

When dz is not an exact differential; that is, when

$$\left(\frac{\partial M}{\partial y}\right)_x \neq \left(\frac{\partial N}{\partial x}\right)_y,$$

z *cannot* be written as a function of x and y alone. Differentials of functions of state are always exact. Another way to find whether a given differential is exact—whether it is the differential of a function of state—is to take the definite integral

(A-5)
$$\int_{x_0,y_0}^{x_1,y_1} dz$$

which specifies a *path* in the x, y plane (Fig. A-2). If z is a function of state, it will be uniquely defined in terms of the values of x and y,

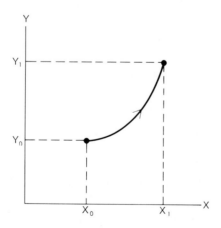

Fig. A-2.

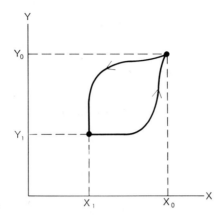

Fig. A-3. Closed path used to calculate integrals of the form $\oint dz$.

and the increment specified by integral A-5 is simply given by the difference between the value of z at the point x_1,y_1 and its value at the point x_0, y_0:

(A-6)
$$\int_{x_0 y_0}^{x_1 y_1} dz = z(x_1, y_1) - z(x_0, y_0)$$

In particular, if the path starts and ends at the same point x_0, y_0 as exemplified in Fig. A-3 the integral will be

$$z(x_0, y_0) - z(x_0, y_0) = 0.$$

This form of integral is called a closed integral and is denoted by

$$\oint dz.$$

A function of state, then, has the following three properties:

(1) $$\oint dz = 0$$

(2) $$z = z(x, y)$$

(3) $$dz = M\,dx + N\,dy$$

$$\left(\frac{\partial M}{\partial y}\right)_x = \left(\frac{\partial N}{\partial x}\right)_y.$$

Any one of these three properties serves to identify z as a function of state.

A-3. First and Second Laws of Thermodynamics

The First Law introduces a function of state, the internal energy E, whose change is given by

$$\Delta E = Q_{in} - W_{out}.$$

For an infinitesimal change,

$$dE = dQ_{rev} - dW_{rev}.$$

Since the reversible pressure-volume work is $dW = P\ dV$ and the reversible heat added is given, by the Second Law, as

$$dQ_{rev} = T\ dS.$$

it follows that

(A-7) $$dE = T\ dS - P\ dV.$$

This expression will be true for either a reversible or an irreversible process, because dE is an exact differential. As before, E has the following properties

(1) $$\oint dE = 0$$

(2) $$\left(\frac{\partial T}{\partial V}\right)_S = -\left(\frac{\partial P}{\partial S}\right)_V$$

(3) $$E = E\ (S,\ V).$$

If there are other work terms (terms other than P, V work), they should also be included in the differential of E. In general

(A-8) $$dE = T\ dS - P\ dV + \Psi\ dq + \mu\ dn$$

A-4. Other Energylike State Functions

Three new functions can be generated from E by applying a special trick called a *Legendre Transformation*. These functions are

the enthalpy, $H = E + PV$;

the Gibbs free energy, $G = E + PV - TS$; and

the Helmoltz free energy, $A = E - TS.$

In differential form they become

$$dH = dE + d(PV) = T\ dS + V\ dP,$$

$$dG = dE + d(PV) - d(TS) = V\ dP - S\ dT,$$

and $$dA = dE - d(TS) = -P\ dV - S\ dT.$$

All these differential forms have been written for a closed system in which only $p\,dV$ work is done; in other cases the appropriate work terms have to be included in dE.

A-5. Explicit Form of the Chemical Potential

When the work expression

$$dW = P\,dV - \mu\,dn$$

is introduced in the differential expression of the First Law, the resultant energy differential is given by

$$dE = T\,dS - P\,dV + \mu\,dn.$$

Furthermore, we previously wrote as the definition of dG

$$dG = dE + d(PV) - d(TS).$$

Replacing dE with its value as given by equation, we obtain

$$dG = T\,dS - P\,dV + \mu\,dn + P\,dV - T\,dS$$

$$dG = \mu\,dn + V\,dP - S\,dT.$$

Since G is a function of state, dG is an exact differential; it then follows that the conditions

(A-9)
$$\left(\frac{\partial \mu}{\partial P}\right)_{T,n} = \left(\frac{\partial V}{\partial n}\right)_{T,P} \quad \text{and}$$

(A-10)
$$\left(\frac{\partial \mu}{\partial T}\right)_{n,P} = -\left(\frac{\partial S}{\partial n}\right)_{P,T}$$

hold. The subscripts outside the parentheses indicate the variables that are held constant in the given differentiation step.

The quantities $(\partial V/\partial n)P,T$ and $(\partial S/\partial n)P,T$ are the partial molar volume and the partial molar entropy, respectively, and are usually denoted by $\overline{V}$ and $\overline{S}$. We integrate Eq. A-6 with respect to P, keeping T and n constant. The equation then reduces to

(A-11)
$$\mu = \int d\mu = \int \overline{V}\,dP + C(T, n)$$

which gives
$$\mu = \mu = \overline{V}P + C(T, P).$$

We can also integrate Eq. A-7 to obtain

$$\mu = \int \overline{S}\,dT + Z\,(P, n)$$

$$= ST + Z\,(P, n).$$

Since the two values of the chemical potential must be the same, the general formula must have the form

$$\mu = \overline{S}T + \overline{V}P + f(n),$$

in which $f(n)$ is only a function of the number of moles in the system and is to be determined. In general, $f(n)$ will vary from system to system. In "ideal" systems $f(n)$ is given by

$$f(n) = RT \ln c$$

for ideal *solutions* and

$$f(n) = RT \ln \overline{P}$$

for ideal *gases*. The chemical potential of a species in an ideal solution is, then,

$$\mu = \overline{S}T + \overline{V}P + RT \ln c.$$

A-6. Differential Definition of the Chemical Potential

Since

$$dG = \left(\frac{\partial G}{\partial n}\right)_{P,T} dn + \left(\frac{\partial G}{\partial P}\right)_{n,T} dP + \left(\frac{\partial G}{\partial T}\right)_{P,n} dT$$

and since this expression is equal to

$$dG = \mu \, dn + V \, dP - S \, dT$$

it is clear that

(A-12)
$$\mu = \left(\frac{\partial G}{\partial n}\right)_{P,T}.$$

That is, the chemical potential is the change in free energy with an increase in the number of moles and when all other variables are held constant.

A-7. The Electrochemical Potential

The transfer of charge in biological systems is always accompanied by the transfer of mass because ions are charges attached to atoms or molecules. Therefore, we can write the charge increment in terms of the increment in mass:

(A-13)
$$dq = (\text{charge per mole})dn.$$

The charge per mole is in turn given by the product of the number of equivalents per mole or charges per ion, z, times the faraday, F, which gives the charge per mole and is equal to $F = 96,500$ coulombs/mole. Equation A-13 then transforms to

(A-14) $$dq = zF \, dn.$$

The number z has to be included with its proper sign depending on whether the species are positively or negatively charged. The new expression for dE is, according to Eqs. A-5 and A-14,

$$dE = T \, dS - P \, dV + \Psi \, dq + \mu \, dn$$

(A-15) $$= T \, dS - P \, dV + \Psi zF \, dn + \mu \, dn.$$

This equation can be simplified to

$$dE = T \, dS - P \, dV + \overset{\sim}{\mu} \, dn$$

by defining the electrochemical potential

(A-16) $$\overset{\sim}{\mu} = \mu + \Psi zF.$$

A-8. Chemical Potentials in Equilibrium and Away from Equilibrium

For motion in one direction—namely, in the x direction—Eq. 7.3

$$\frac{F}{n} = -\frac{\Delta\mu}{d}$$

has the differential form

(A-17) $$\frac{F}{n} = -\frac{d\mu}{dx}$$

Thus, the force per unit mole on a given uncharged species is the spatial derivative of μ (or $\overset{\sim}{\mu}$ for a charged species). At equilibrium, the net force is zero and

(A-18) $$-\frac{d\mu}{dx} = 0$$

We should point out that Eq. A-17 is valid only for systems in which energy is conserved (conservative systems). Otherwise, the force cannot be found from a potential.

A-9. External Forces and Steady State

If a force F is acting on a particle of mass m, the equation of motion is $F = m \times d^2v/dt^2$. Let the external force F be composed of two forces: a constant driving force, f, and a frictional force directed in the direction opposite to the motion. Furthermore, assume that the frictional force is proportional to the velocity of the molecule. Then, the total force can be written as

(A-19) $$F = f - f_{friction} = ma \qquad \text{or}$$

(A-20) $$f = ma + kv,$$

where K is a positive constant. If the acceleration dies out fast, this equation becomes $F = k \times v$, so the force is also linearly related to the velocity v. In order to check this result and in order to get some quantitative estimate of how fast the acceleration must decay, Eq. A-20 can be put in differential form. When this is done, it becomes

(A-21) $$f = m \frac{dv}{dt} + kv$$

which has the solution

(A-22) $$v_{(t)} = \frac{f}{k} [1 - e^{-(k/m)t}],$$

for a molecule starting from absolute rest; that is, $v(0) = 0$. If (kt/m) is much larger than 1, the solution looks like

(A-23) $$v = \frac{f}{k},$$

that is, the velocity is a constant independent of time. This will happen for $t \gg m/k$ (seconds). Thus, if a measurement is made for a time larger than m/k, the velocity will essentially be a constant.

The constant k is called the coefficient of friction; sometimes it is convenient to use another constant defined as the inverse of k, $1/k$. This new constant is called the mobility ω. Introducing the mobility ω, Eq. A-23 becomes

$$v = \omega f.$$

ω can then be thought of as the velocity produced by a unit force.

A-10. Chemical Potential Outside Equilibrium: Fick's Law

The steady flow of solute per unit area is given by

$$(\text{A-24}) \qquad\qquad J_s = c_s \omega f,$$

if the flow of solute and water are assumed to be independent. The force f is, as before, the derivative of the chemical potential in the direction of motion (for simplicity, we have considered only one-dimensional flow). The flow of Eq. A-24 becomes

$$(\text{A-25}) \qquad\qquad J_s = -\, c_s \omega \, \frac{d\mu_s}{dx}$$

and, replacing the differential $d\mu_s$ by its explicit value at constant T, P, and ψ, this equation yields:

$$J = (-c) \, \omega RT \, \frac{d \ \ln \ c}{dx} = (-c) \ \frac{\omega RT}{c} \ \frac{dc}{dx}$$

$$(\text{A-26}) \qquad\qquad J = -\, \omega RT \, \frac{dc}{dx}.$$

This result is Fick's Law of Diffusion, if the concentration inside the membrane changes linearly (Fig. A-4). Equation A-26 can be integrated across a membrane of thickness d to yield

$$\int_0^d J \, dx = -\omega RT \int_{c(1)}^{c(2)} dc \qquad \text{or}$$

$$(\text{A-27}) \qquad\qquad J = \frac{\omega RT}{d} (c_2 - c_1),$$

in which c_2 and c_1 are the concentrations in the two compartments. If c_2 is larger than c_1, the flow will go in the direction of negative x, as expected. Equation A-27 is usually rewritten in the form

$$(\text{A-28}) \qquad\qquad \frac{dn}{dt} = \frac{DA}{d} (c_2 - c_1),$$

in which we have defined the diffusion coefficient

$$D = -\omega RT,$$

and the expression is given in terms of the *total* number of moles that go through the partition in unit time.

A-11. Chemical Reactions in the Steady State

Given a reaction of the type

(A-29) $aA + bB + cC \ldots \longrightarrow lL + mM + nN \ldots$

one rarely finds that a, b, c, $\ldots$, moles of the reactants have disappeared to yield l, m, n, $\ldots$, moles of products. Instead, different numbers are found although, according to the rules of elementary chemistry, A, B, C, $\ldots$, will always react in the proportions given by Eq. A-29; that is, the amount of reactants and products will always be given by the equation above, appropriately multiplied by some factor that we call the degree of advancement and denote by $\Delta\zeta$. Equation A-29 then becomes

(A-30) $\Delta\zeta\, aA + \Delta\zeta\, bB + \Delta\zeta\, cC \ldots \longrightarrow \Delta\zeta\, lL + \Delta\zeta\, mM + \ldots$

The change in free energy is given by

$$\Delta G_{T,P} = -a\mu_A\, \Delta\zeta - b\mu_B\, \Delta\zeta - c\mu_C\, \Delta\zeta + \ldots + l\mu_L\, \Delta\zeta + m\mu_M\, \Delta\zeta + \ldots$$

$$= \Sigma_i \nu_i\, \mu_i\, \Delta\zeta.$$
(A-31)

In this formula, ν_i represents the coefficients a, b, c, $\ldots$, l, m, $\ldots$ with their appropriate signs. The summation is taken over all reactants and products. The degree of advancement, $\Delta\zeta$, can be taken outside the summation sign, so Eq. A-31 can be written as

(A-32) $\Delta G = \Sigma_i \nu_i \mu_i \times \Delta\zeta.$

When $\Delta\zeta$ is very small, Eq. A-32 can be expressed in differential form to yield

$$dG = (\Sigma_i \nu_i \mu_i)\, d\zeta.$$

In the steady state, chemical potentials are independent of time; then we can write for dG/dt:

(A-33) $\dfrac{dG}{dt} = \Sigma_i \nu_i \mu_i \dfrac{d\zeta}{dt}.$

The derivative $d\zeta/dt$ gives the number of moles of the reaction that are produced per second. Denoting the summation in parenthesis by $-A$ and $d\zeta/dt$ by J_{chem}, Eq. A-33 simplifies to

(A-34) $\dfrac{dG}{dt} = -AJ_{chem}.$

This is a useful equation for describing chemical processes in the steady state; it will be shown that J_{chem} is analogous to a flow, and that A represents the driving force in the chemical reaction.

A-12. The Internal Production of Entropy

The differential change in free energy for any process is given by

$$dG = d(E) + d(PV) - d(TS),$$

which at constant temperature and pressure becomes

$$dG_{P,T} = dE + P\,dV - T\,dS.$$

According to the First Law of Thermodynamics

(A-35) $$dE = dS - dW,$$

in which $$dG = TdS - Td_iS$$

and $$dW = P\,dV + dW'.$$

Equation A-35 can be rewritten as

$$dG_{p,T} = T\,dS - T\,d_iS - P\,dV - dW' + P\,dV - T\,dS$$

(A-36) $$= -T\,d_iS - dW'.$$

If no "useful" work dW' is done on or by the environment, the change in free energy is

(A-37) $$dG_{T,p} = -T\,d_iS,$$

which can be differentiated with respect to time to give

(A-38) $$\left(\frac{dG}{dt}\right)_{T,P} = -\frac{T\,d_iS}{dt}.$$

The quantity $T\,d_iS/dt$ is called the *dissipation function* and it is denoted by Φ.

We previously derived for the time change in free energy in a chemical reaction

$$\frac{dG_{T,p}}{dt} = -J_{chem} \times A,$$

in which A is the chemical affinity, and in which J_{chem}, the flow of chemical reaction, is the time derivative of the degree of advancement. Similarly, we can derive from Eq. 3-15, the expression for the free-energy change per unit time when matter flows between two compartments,

$$\left(\frac{dG}{dt}\right)_{T,P} = -\,J\,\Delta\mu,$$

or, in a continuous solution,

$$\left(\frac{dG}{dt}\right)_{T,P} = J \times \frac{d\mu}{dx}.$$

The corresponding dissipations for the three equations above are, from Eq. A-38,

$$\Phi_1 = J_{\text{chem}} \times A,$$

$$\Phi_2 = J\,\Delta\mu,$$

and

$$\Phi_3 = J\,\frac{-d\mu}{dx},$$

respectively. If two chemical reactions are taking place at the same time, we can add up their individual dissipations:

$$\Phi = J_{\text{chem}}^{(1)}\,A^{(1)} + J_{\text{chem}}^{(2)}\,A^{(2)}.$$

Similarly, if two substances are diffusing at the same time

$$\Phi = J_1\,\Delta\mu_1 + J_2\,\Delta\mu_2,$$

or

$$\Phi = J_1\left(\frac{-d\mu_1}{dx}\right) + J_2\left(\frac{-d\mu_2}{dx}\right).$$

In general, we have to add up the individual dissipations of all processes involved and the final dissipation function will have the form

$$\Phi = J_1 X_1 + J_2 X_2 + \ldots + J_N X_N,$$

in which conjugate flows and forces have the same subscript. In the special case of a two force–two flow system, the dissipation becomes

$$\Phi = J_1 X_1 + J_2 X_2.$$

Since entropy must increase with time in a natural process,

$$T\frac{d_i S}{dt} > 0$$

and $$\Phi > 0.$$

This inequality tells us that the *overall* dissipation function, $X_1 J_1 + X_2 J_2$, must be positive, but it does not rule out the possibility of having a negative term, say $X_2 J_2$, as long as

$$X_1 J_1 + X_2 J_2 > 0$$

or $$X_1 J_1 > -X_2 J_2.$$

Then $$\frac{X_2 J_2}{X_1 J_1} < 1;$$

that is, the efficiency must be less than one.

A-13. Setting Up the Proper Equations

Onsager showed that given a dissipation function

$$\Phi = J_1 X_1 + J_2 X_2$$

and assuming that X_1 is independent of X_2 (or J_1 is independent of J_2), the following linear equations can be written

$$J_1 = A_{11} X_1 + A_{12} X_2$$
$$J_2 = A_{21} X_1 + A_{22} X_2.$$

Furthermore, $A_{12} = A_{21}$. These equations are called phenomenological equations. In some cases the dissipation function

$$\Phi = J_1 X_1 + J_2 X_2,$$

in which the primed variables are new flows and forces. These new flows and forces are related by the equations

$$J_1' = A_{11}' X_1' + A_{12}' X_2'$$
$$J_2' = A_{21}' X_1' + A_{22}' X_2'$$

and $A_{12}' = A_{21}'$ if certain mathematical restrictions are met. We shall now give a few examples already considered in the test from a different point of view.

A-14. Diffusion of a Solute in Water

We previously derived Fick's law by assuming that the only flow was the solute flow and that the only driving force was the spatial gradient of solute electrochemical potential. Actually, there are two flows, the flow of water and the flow of solute, but they are not independent. Their relationship is found from the following (Gibbs-Duhem) equation:

(A-39)
$$c_s \frac{-d\mu_s}{dx} + c_w \frac{-d\mu_w}{dx} = 0$$

in which the subscripts "s" and "w" refer to the solute and water, respectively. The forces that drive water and solute are then related according to

(A-40)
$$\frac{-d\mu_w}{dx} = -\frac{c_s}{c_w} \frac{-d\mu_s}{dx}.$$

The total dissipation function is given by

$$\Phi = \frac{T\, d_iS}{dt} = J_w \frac{-d\mu_w}{dx} + J_s \frac{-d\mu_s}{dx},$$

which, after introducing Eq. A-40, becomes

$$\Phi = \left(J_s - J_w \frac{c_s}{c_w} \right) \frac{-d\mu_s}{dx}.$$

This dissipation function indicates that there is one *independent* force, the gradient of solute, and one *independent* flow, the diffusional flow,

$$J_d = J_s - J_w \frac{c_s}{c_w}.$$

If we replace J_s by $c_s v_s$ and J_w by $c_w v_w$, we find the diffusional flow is given by

$$J_d = c_s(v_s - v_w).$$

The actual diffusional flow, then, is not the flow of solute but the *relative* motion of solute with respect to water.

Since in this particular case there is only *one* flow and *one* force, the phenomenological equations reduce to one:

$$J_d = D' \frac{-d\mu_s}{dx},$$

in which D' approaches the previously found diffusion constant $D = \omega RT$ in dilute solutions.

A-15. Passage of Solute and Water Across a Membrane

The dissipation function for this case can be written as

$$\Phi = J_w\,\Delta\mu_w + J_s\,\Delta\mu_s.$$

Introducing the explicit potential differences in dilute solutions, it is possible to show (Prob. A-4) that the dissipation function transforms to

(A-41) $$\Phi = (J_w\overline{V}_w + J_s\overline{V}_s)\,\Delta P + \left(\frac{J_s}{\overline{C}_s} - \overline{V}_w\,J_w\right)\Delta\pi$$

in which $\overline{C}_s$ is the *average concentration of solute in the membrane*; $\Delta\pi$ is the *ideal osmotic pressure*, $RT\,\Delta C_s$; $\overline{V}_w$ and V_s are the *partial molar volumes* of water and solute, respectively; and ΔP is the *hydrostatic pressure* difference across the membrane.

According to Onsager, the new dissipation function is also the product of conjugate flows and forces. If we rewrite Eq. A-41 in a "shorthand" fashion, it can be abbreviated to

$$\Phi = J_v\,\Delta P + J_D\,\Delta\pi$$

with $$J_v = (J_w V_w + J_s V_s),$$

and $$J_D = \frac{J_s}{\overline{C}_s} - \overline{V}_w\,J_w\,.$$

Nonequilibrium thermodynamics says that J_v and J_D are the flows conjugate to the forces ΔP and $\Delta\pi$ respectively, and the phenomenological equations associated with the new dissipation are

$$J_v = L_p\,\Delta P + L_{pD}\,\Delta\pi$$

$$J_D = L_{DP}\,\Delta P + L_D\,\Delta\pi,$$

in which L_p, L_{pD}, L_{Dp}, L_D are numbers that depend on the given system and $L_{Dp} = L_{pD}$. The flow J_v is called the *volume flow*, and, for dilute solutions, it is about equal to the flow of water, J_w. The flow J_D is the *diffusional* flow of solute, and it is approximately equal to the relative velocity of the solute with respect to the water; that is,

$$J_D = v_s - v_w\,.$$

These equations are completely equivalent to those found before in Chapter 7 using different methods.

Problems

A-1. (a) Write the First Law of Thermodynamics for the expansion of an ideal gas. (The energy of an ideal gas is proportional only to the temperature, so an infinitesimal change in energy, dE, can be expressed as $dE = C_V \, dT$ in which C_V is a constant.)

(b) Utilize the result found in (a) to give an expression for the ratio dQ/T in terms of C_V, R, V, T, dT, and dV.

(c) Show that $dS = dQ/T$ is an exact differential.

A-2. (a) Calculate the change in entropy, ΔS, for the following path: increase in temperature, from T_0 to T_1 at constant volume V_0, followed by expansion to V_1 at constant temperature T_1.

(b) Repeat the calculation when the expansion at constant temperature T_0 precedes the temperature increase. Compare with (a).

(c) Calculate the heat added for the paths of (a) and (b). Discuss.

A-3. (a) The Helmoltz free energy, A, is defined by

$$A = ST - E.$$

Give the differential dA in terms of dT, dV, S, and P.

(b) Use your result above to prove that

$$\left(\frac{\partial S}{\partial V}\right)_T = \left(\frac{\partial P}{\partial T}\right)_V.$$

(c) Use the First Law of Thermodynamics, the equation of state of an ideal gas, and the result of (b) to prove that a change in volume at constant temperature does not change the internal energy of an ideal gas.

A-4. Show that two adiabats cannot cross in the P-V plane.

A-5. Calculate the work done when a gas expands adiabatically from a volume V_0 to a volume $2V_0$.

GLOSSARY

Action potential. Electrical disturbance transmitted along the nerve axon which according to modern theories is caused by a transient change in the Na^+ and K^+ permeabilities of the axon membrane.

Activation energy. Free energy required, in addition to the equilibrium free-energy change, for a reaction or process to take place. The activation energy depends on the structure and geometry of the system.

Activity coefficients, γ. Multiplicative correction factors that account for the non-ideal behavior of a solute. By taking the effective concentration of a solute, c, as γC, one can pretend that the solution is ideal and utilize thermodynamic formulas. The resultant effective concentration, γC is called the activity of the solute, a.

Adenosine tri-phosphate. Molecule which can release one or two phosphate moieties with a concomitant release of free energy. As the phosphate groups that appear during the hydrolysis of ATP can be transferred to other molecules, the free energy of ATP hydrolysis may be coupled to other reactions. Thus, ATP serves as an intermediate step in the transfer of energy from reactions which release large amounts of negative free energy and those which require relatively small amounts. ATP is incorrectly called a "high energy compound."

Bioenergetics. In the present context, the study of biological energy transactions utilizing physicochemical methods. A more restricted view held by many biochemists considers bioenergetics as the study of ATP generation and utilization.

Biosphere. Total portion of the earth where life exists.

Bit. Short for binary digit and unit of measurement for information content. One bit is the number of binary digits required to code for a decision step in which the possible outcomes have equal probabilities.

Bonds, chemical. Forces that hold atoms and molecules together. In covalent bonding, the strongest chemical bond, pairs of electrons are shared between two atoms. There are also weaker electrical interactions, such as Van der Waals and ionic bonding, which can hold atoms and molecules together.

Bulk flow (Poiseulle). Steady motion of a fluid through a tube across which a hydrostatic pressure difference has been established.

Chemical potential, μ_i. Change in free energy per unit mole required to add an uncharged species. The chemical potential measures the tendency of the species to leave the region of space or the state in which it is. Chemical potentials can only be measured relative to an arbitrarily set reference value.

Concentration, c_i. Amount of a given substance per unit volume. Throughout this text we have denoted all concentrations in Molar units (M) by specifying the number of moles of that substance (6×10^{23} molecules) per liter of solution.

Cyclic process. A process that returns to an initial state.

Diffusion. Molecular motion imparted by the random thermal motions of water (or other solvent) on the dissolved solute molecules, which tends to equate the concentration of the solute in different parts of a connected system.

Diffusion potential. Electrical potential differences which appear when an electrolyte moves down a concentration gradient. These potentials are caused by the difference in mobility between anion and cation, so that a polarization effect is obtained as there is an excess of the fastest ion in the region of lower concentration.

Dissipation function, ϕ. The time rate of change of internal entropy production multiplied by the temperature—in a system at constant temperature. This quantity is always larger than or equal to zero.

Donnan equilibrium. Ionic equilibrium across a membrane permeable to anion and cation when the membrane (or other partition) separates a compartment with fixed charges from a compartment containing only the electrolyte. In the context of this book we have considered that the fixed charges are on proteins which cannot go across the partition.

Ecosystems. A thermodynamic system that includes a group of interacting organisms and their environment.

Electrical potential, ψ. Intensive variable which measures the work needed to bring an electrical charge from a reference region to that region of space being considered.

Electrochemical potential, μ_i. The change in free energy per unit mole of a given charged species added.

Entropy, S. Thermodynamic function of state which measures the likelihood that a given transition will take place. Entropy units are energy/degree mole.

Equilibrium. State in which the properties of a system can be completely determined by a set of external variables. At equilibrium there is no flow of energy across the boundary that separates the system from the environment, and all the physical properties remain constant in time.

Extensive variables. Properties of a system whose values depend on the amount of system considered. Examples are energy, mass, charge, and volume. The ratio of two extensive variables is an intensive variable.

First Law of Thermodynamics. Statement of the law of energy conservation for closed systems. Mathematically, the statement that energy is a function of state whose increment equals the heat added to the system minus the work done by the system on the environment.

Free energy, G. Function of state whose change, at constant temperature and pressure, gives the maximum amount of work a system can perform reversibly under these conditions. Free energy is defined as $G = E + PV$, and it always decreases in a spontaneous, or natural, process.

Functions of state. Quantities which can be specified at equilibrium in terms of measurable variables such as temperature, pressure, etc. The simplest functions of state are the measurable quantities; others are calculated from these—e.g., energy and entropy.

Genetic code. The set of triplet RNA bases each of which codes for one amino acid, a start, or a termination signal required to assemble a protein. The genetic code is usually given in terms of RNA, rather than DNA triplets.

Heat, Q. Form of energy which reflects the internal molecular motions of a system.

Hydrostatic pressure. Pressure applied to an incompressible fluid which transmits with the same value throughout the system.

Information content, H. Measure of the degree of organization needed to transmit a specific set of messages having certain likelihoods of appearance. If all messages are equally probable, H is given by $H = \ln_2 p$, in which p is the probability that a given message will be transmitted. If the probabilities are different, H is given by $H = \Sigma p_i \ln_2 p_i$.

Information theory. Mathematical model that considers how the entropy changes involved in message transmission can be measured.

Intensive variables. Quantities which, at equilibrium, do not depend on the amount of a system considered. Examples are density, concentration, temperature, and pressure.

Internal entropy production, $\Delta_i S$. Entropy generated inside a system during an irreversible process. During a reversible transition, $\Delta_i S$ is zero.

Irreversible process. A process which cannot be returned to its initial state by doing the same amount of work that brought about the irreversible transition. All natural processes are irreversible.

Kinetic energy, K.E. Energy of motion. For a body of mass m moving at constant velocity v, this energy is given by $E = \frac{1}{2} mv^2$.

Kinetics. The analysis of processes at the level of molecular motions and collisions. As opposed to thermodynamics, kinetics is a microscopic theory.

Molecular fluctuations. Random molecular motions which make a system depart from the average values of their macroscopic properties.

Nernst potential. Electrical potential which, at equilibrium, appears in a system composed of two compartments separated by a membrane permeable to one ion.

Nonequilibrium thermodynamics. Modern thermodynamic theory that considers steady-state processes rather than transitions between equilibrium states. The main difference between classical thermodynamics and nonequilibrium thermodynamics is the introduction of the variable time in the latter.

Osmotic pressure, $\Delta\pi$. Pressure developed across a semipermeable membrane which separates two compartments in which the impermeant solute is maintained at different concentrations. If the concentration difference of the solute is Δc, the ideal osmotic pressure is $\Delta\pi = RT\,\Delta C$.

Partial molar volume, $\overline{V}_i$. The increase in volume of a system per mole of the species i added. The total volume of the system in which there are n species present is given by $V = \overline{V}_1 n_1 + \overline{V}_2 n_2 + \ldots \overline{V}_n n_n$.

Potential energy. Nonkinetic energy stored in a system in a form available to do work.

Power. Rate of addition or removal of energy from a system. The most common unit of power is the Watt = Joule/sec.

Pressure, P. Force applied per unit area, $P = F/A$. A common unit is the atmosphere which is the pressure of a column of mercury 760 mm high.

Resting potentials. Equilibrium potential differences of Nernst origin found across the membranes of most living cells.

Reversible process. Thermodynamic change which can be brought back to its initial state by doing the same amount of work that brought about the original change.

Second Law of Thermodynamics. Any of the statements which, directly or indirectly, specify that the total entropy of a system plus its surroundings must increase during an irreversible process.

Statistical mechanics. Part of physics which studies average microscopic interactions among many particles, atoms, or molecules without considering individual motions. The average quantities obtained by statistical mechanics are the macroscopic, measurable quantities.

Thermodynamic efficiency. Ratio of the output work to the heat input. More generally, ratio of output work vs. energy input.

Thermodynamic systems. Any region that can be enclosed by an imaginary boundary and described in terms of thermodynamic variables.

Thermodynamics. Part of physics and chemistry which studies functions of state and their changes between equilibrium states.

Transducers. Any biological or inanimate machine that transforms one kind of energy into another.

Useful work, W'. Term currently used to denote any biochemical work that is not pressure-volume work. If W is the total work expression, the useful work is calculated from the difference $W' = W - P\,\Delta V$.

Work, W. The mechanical form of energy. For a constant force, F, it is simply the product of the force times the displacement in the direction of the force.

SOLUTIONS TO PROBLEMS

Chapter 2

2-1. Our previous discussion indicated that $Q_{\text{initial}} - Q_{\text{final}}$ is not a function of state, in general, because the work done depends on the path between initial and final states. If, however, no work is done in the transition and only heat is added or removed, the First Law reduces to

$$Q_{\text{initial}} - Q_{\text{final}} = E_{\text{initial}} - E_{\text{final}}$$

Since the change in energy is independent of the path, the heat change will be too.

The same considerations apply to the work done on or by the system when it is adiabatic.

2-2. The basic assumption is that *all* of the mechanical—or electrical—energy can be transformed into heat; this is true only if the energy conversion process is 100% efficient. As we shall see in Chapter 10, it is impossible to make a machine which is that efficient *unless* energy is transformed completely into heat without obtaining work as an output. Since these are the conditions of Joule's experiment, the measurement is valid.

2-3. Equilibrium thermodynamics considers, by definition, systems in which all exchanges of energy and matter with the environment have ceased. No flows take place across the boundary of the system. Living biological systems depend on these exchanges to keep their metabolic machinery running. The only time an organism reaches equilibrium is at death. The justification for considering equilibrium thermodynamics at all in this context is that living organisms are open systems which reach a dynamic "equilibrium" state, the steady state, in which rates of all processes are constant in time.

We indicate in later chapters how steady-state thermodynamics is based on equilibrium thermodynamics.

Chapter 3

3-1. The expression $\mu = RT \ln c$ has no meaning by itself; it becomes meaningful only when a reference potential, μ^0, is established relative to which the "height" of the potential is measured. Since

$$\mu - \mu^0 = RT \ln c - RT \ln c_0,$$

in which c_0 is the reference concentration, the logarithmic forms are combined to give

$$\mu - \mu^0 = RT \ln \frac{c}{c_0}.$$

The expression is meaningful because the logarithm of a concentration *ratio* is simply the logarithm of a number.

3-2. (*a*) Consider a transition $1 \rightarrow 2$. The enthalpy change is given by

$$H_2 - H_1 = (E_2 + P_2 V_2) - (E_1 - P_1 V_1)$$

Since the initial and final quantities are uniquely defined in terms of the initial and final states, the enthalpy change is also uniquely determined once E_1, E_2, P_1, P_2, V_1, and V_2 are specified.

(*b*) We can rewrite the change given in (*a*) as

$$\Delta H = \Delta E + P(V_2 - V_1)$$

$$= \Delta E + P \, \Delta V,$$

at constant pressure. According to the First Law,

$$\Delta E = Q_{added} - W_{output}$$

or, for work of expansion,

$$\Delta E = Q_{added} - P \, \Delta V.$$

Introducing this expression in H, we obtain

$$\Delta H = (Q_{added} - P \, \Delta V) + P \, \Delta V$$

$$= (Q_{added})$$

The enthalpy increase at constant pressure is, then, equal to the heat added to the closed system.

3-3. The change in entropy in the system can be calculated in two parts:

$$\Delta S_A = \frac{-Q}{T_A}$$

and

$$\Delta S_B = \frac{Q}{T_B};$$

in which the minus sign on the first equation indicates that heat leaves—hence, entropy decreases. The total change in entropy then becomes

$$\Delta S_{total} = \Delta S_A + S_B$$

$$= \frac{Q}{T_B} - \frac{Q}{T_A}$$

$$= Q \frac{(T_A - T_B)}{T_A T_B}$$

Since the total entropy must increase this last expression must be positive, which can only occur if T_A is larger than T_B in accordance with Clausius postulate.

3-4. The change in free energy for a substance moving between two compartments, 1 and 2, given by

$$\Delta G = (\mu_2 - \mu_1)\Delta n.$$

Or, for 1 mole,

$$\Delta G = \mu_2 - \mu_1.$$

According to Eq. 3-25,

$$\Delta G = \mu_2 - \mu_1 = \overline{V} \Delta P - \overline{S} \Delta T + RT \ln \frac{c_2}{c_1} + zF \Delta \Psi .$$

If the molecules are uncharged and the two regions are at the same T,P, we have $\Delta \Psi = 0$, $\Delta T = 0$ and $\Delta P = 0$. At equilibrium, $\Delta G = 0$,

$$0 = RT \ln \frac{c_2}{c_1}.$$

In order for the logarithm to be zero, the ratio c_2/c_1 must equal 1, or

$$c_2 = c_1.$$

3-5. The "exit" of mole of A changes the free energy of "compartment A" by

$$\Delta G_A = - (1) \mu_A$$

while the "entry of B" changes the free energy of "compartment B" by

$$\Delta G_B = + (1) \mu_B.$$

The total change in free energy is

$$\Delta_G = \mu_B - \mu_A .$$

At constant temperature and pressure the chemical potentials of A and B can be written as

$$\mu_B = \mu_B^0 + RT \ln C_B$$

and

$$\mu_A = \mu_A^0 + RT \ln C_A.$$

The change in free energy is, then,

$$\Delta G = (\mu_A^0 - \mu_B^0) + RT \ln \frac{C_B}{C_A}$$

or using the terms defined in the problem,

$$\Delta G = \Delta G^0 + RT \ln K_{eq}.$$

3-6. (a) At equilibrium $\Delta G = 0$.
(b) Since $0 = \Delta G^0 + RT \ln K_{eq}$,

$$\Delta G^0 = - RT \ln K_{eq}.$$

This result can also be rewritten, equivalently, as,

$$K_{eq} = e^{-\Delta G^0/RT}$$

a result we shall use in Chapter 5.

(c) If ΔG^0 is negative, K_{eq} will be larger than 1 and c_B will be larger than c_A. If ΔG^0 is positive, the logarithm is negative and K_{eq} is smaller than 1; in this case c_A is larger than c_B.

3-7. By using the same steps as in Prob. 3-5 we obtain

$$\Delta G = m\mu_M + n\mu_n + u\mu_u - a\mu_A - b\mu_B - c\mu_c,$$

which, after rearrangement and introduction of the explicit form of the chemical potentials at constant T, P becomes

$$\Delta G = \Delta G^0 + RT \ln K_{eq};$$

in which $\quad\quad \Delta G^0 = m\mu_M^0 + n\mu_N^0 + u\mu_u^0 - a\mu_A^0 - b\mu_B^0 - c\mu_C^0$

and $\quad\quad\quad\quad K_{eq} = \dfrac{[M]^m [N]^n [U]^u}{[A]^a [B]^b [C]^c}$

3-8. The free-energy change is given by

$$\Delta G = E_2 - E_1 + P_2 V_2 - P_1 V_1 - T_2 S_2 - T_1 S_1$$

$$\Delta G = \Delta E + \Delta(PV) - \Delta(TS).$$

At constant T, P, we obtain

$$\Delta G_{T,P} = \Delta E + P \Delta V - T \Delta S.$$

Furthermore, the First Law gives

$$\Delta E = Q - P \Delta V - W'$$

is all work other than $P\Delta V$ work.

The entropy change can be separated into two parts

$$\Delta S = \frac{Q}{T} + \Delta_i S,$$

in which the first term is the entropy increase caused by the entry of heat into the system and the second term the internal entropy increase caused by the irreversible processes. The heat added can then be written as

$$Q = T \Delta S - T \Delta_i S,$$

which gives for the energy change

$$\Delta E = T \Delta S - T \Delta_i S - P \Delta V - W'.$$

Introduction in the expression for ΔG yields, after cancelation of terms with opposite signs,

$$\Delta G_{T,P} = W' - T \Delta_i S.$$

During an irreversible process at constant T, P, in which no work is done on the environment,

$$\Delta G_{T,P} = -T \Delta_i S.$$

Since $\Delta_i S$ is either zero or positive, $\Delta G_{T,P}$ will in general decrease.

3-9. The maximum amount of work will be obtained from a *reversible* process, when $\Delta_i S = 0$, in this case

$$-\Delta G_{T,P} = W'.$$

3-10. In an irreversible process,

$$-(\Delta G_{T,P} - T \Delta_i S) = W'.$$

Since $\Delta_i S > 0$ the work done will be less than the free-energy decrease.

3-11. The term "path" refers to the route between a given initial state and a given final state, where all measurable variables are uniquely defined. There are *many* systems that can go between *the same* initial and final states. When we observe a process taking place in a *specific* system, however, the path for the reversible process has the same initial state as for an irreversible process but a *different* final state. The entropy change for the two processes is then different. But one can always find a different reversible system which has the same initial and final states of the irreversible process considered. The entropy change for the reversible process will be the same as the entropy change in the irreversible processes, but the systems and processes are different.

Chapter 4

4-1. The change in entropy for an expansion of a gas at constant temperature can be found from the First and Second Laws (in incremental form),

$$\Delta E = -P \Delta V + T \Delta S.$$

Since the change in energy for an ideal gas is given by $\Delta E = C_V \, \Delta T$, there is no change in energy at constant temperature. The change in entropy is, then, for a small change,

$$\Delta S = \frac{P}{T} \, \Delta V.$$

The ratio P/T is, according to the equation of state of ideal gases, nR/V. Substitution in the expression above yields

$$\Delta S = nR \, \frac{\Delta V}{V}.$$

As we discussed in the text, for a large change this expression has the form

$$\Delta S = S_2 - S_1 = nR \, \ln \frac{V_2}{V_1}$$

Chapter 5

5-1. (*a*) The minimum amount of work required is the free-energy change needed to pump the solute reversibly,

$$W = \Delta G_{T,P} = n \, \Delta\mu$$

in which $\mu = RT \ln e^2 = 2 \, RT$. The work done per mole is then

$$\frac{W}{n} = 2 \, RT = 2 \, (2 \text{ cal/mole K}) \, (298^\circ \text{ K})$$

$$= 1192 \text{ cal/mole}.$$

(*b*) If the solute is charged, the work per mole is given by

$$\frac{W}{N} = \Delta\mu = RT \ln c_{in}/c_{out} + zF \, (\Psi_{in} - \Psi_{out})$$

$$= 1192 \text{ cal/mole} + (1) \, (96,500) \text{ coul} \, (58 \times 10^{-3}) \text{ volt}$$

$$= 10,577 \text{ joules/mole} = 2536 \text{ cal/mole}.$$

5-2. As ΔG^0 alone does not determine whether a process, in particular a chemical reaction, will occur or not, we must first find an explicit expression for the change in total free energy, ΔG. The total change in free energy for the reaction is

$$\Delta G = \Delta G^0 + RT \ln \frac{[suc]}{[fru] \, [glc]}$$

The reaction will take place in the direction indicated when the total free-energy change is negative, that is,

$$RT \ln \frac{[suc]}{[fru] \, [glc]} < -5,500 \text{ cal/mole}.$$

At a temperature of $300°$ K this implies that

$$\ln \frac{[\text{suc}]}{[\text{fru}]\,[\text{glc}]} < -9.21$$

The reaction will then take place provided that the ratio of the concentration of sucrose to the product of concentrations of fructose and glucose obeys the inequality,

$$\frac{[\text{suc}]}{[\text{fru}]\,[\text{glc}]} < e^{-9.21}$$

or

$$\frac{[\text{suc}]}{[\text{fru}]\,[\text{glc}]} < 10^{-4}$$

5-3. As

$$\Delta G^0 = -RT \ln K_{\text{eq}}$$

we obtain

$$\ln K_{\text{eq}} = \frac{-\Delta G^0}{RT} = \frac{+1{,}500}{RT}\ \text{cal/mole},$$

which, at $T = 300°$ K gives

$$K_{\text{eq}} = e^{2.513} = 12.34.$$

In order for the reaction to occur,

$$RT \ln \frac{[\text{suc}][\text{ADP}]}{[\text{fru}][\text{glc}][\text{ATP}]} < \Delta G^0$$

At $300°$ K,

$$\ln \frac{[\text{suc}][\text{ADP}]}{[\text{fru}][\text{glc}][\text{ATP}]} < 2.513.$$

5-4. Under the best possible conditions, in which the reactions couple completely, 5000/7000 moles of ATP would be required or 0.714 moles.

5-5. As 2 moles are required stoichiometrically, the efficiency is

$$\frac{0.714}{2} \times 100 = 35.7\%.$$

Chapter 6

6-1. According to Eq. 6-11,

$$[\text{K}^+]_2 = [\text{Cl}^-]_2 + [\text{P}]\nu$$

and, dividing both sides by $[\text{K}^+]_1$, we obtain

$$\frac{[\text{K}^+]_2}{[\text{K}^+]_1} = \frac{[\text{Cl}^-]_2}{[\text{K}^+]_1} + \frac{[\text{P}]\nu}{[\text{K}^+]_1}$$

Since, according to Eq. 6-10, $[K^+]_1 = [Cl^+]_1$, we can also write

$$\frac{[K^+]_2}{[K^+]_1} = \frac{[Cl^-]_2}{[Cl^-]_1} + \frac{[P]\nu}{[K^+]_1}$$

and, substituting explicitly for r, we obtain

$$\frac{1}{r} = r + \frac{[P]\nu}{[K^+]_1}$$

or

$$r^2 + \frac{r[P]\nu}{[K^+]_1} - 1 = 0$$

This is a second-degree equation with two solutions:

$$r = \frac{[P]\nu}{2[K^+]_1} \pm \sqrt{\left(\frac{\nu[P]}{2[K^+]_1}\right)^2 + 1}$$

We adopt the solution with positive sign because the negative solution gives an answer with no physical meaning (check what happens in both cases when $\nu[P]$ is very large).

6-2. (a) Since the Donnan effect depends on the expulsion of ions by the protein charges, an uncharged protein should have no effect on the ion distribution. Furthermore, the membranes or partitions considered are permeable to all ions so, at equilibrium, there will be no concentration difference across the partition. The Donnan ratio, r, is then 1. This is verified by placing $\nu = 0$ in the Donnan formula.

(b) When $[K^+]_1$ is very large, the electrostatic forces that tend to expel the positive ions from the protein compartment balance the electrostatic forces which tend to push the K^+ into the same region. The Donnan ratio is 1 (this can be verified by placing $[K^+]_1 = \infty$ in the Donnan formula).

(c) In both cases the electrical potential difference across the partition is zero.

6-3. The relevant variables are:

$$\nu = +8$$

$$[P] = 10 \times 10^{-3} M$$

$$[K^+]_1 = 300 \times 10^{-3} M.$$

(a) We wish to find $[K^+]_2$. The Donnan ratio is

$$r = \frac{8 \times 10 \times 10^{-3}}{2 \times 300 \times 10^{-3}} + \sqrt{\left(\frac{8 \times 10 \times 10^{-3}}{2 \times 300 \times 10^{-3}}\right)^2 + 1}$$

$$= 1.138$$

The concentration of $[K_2^+]$ is, then,

$$[K^+]_2 = \frac{[K^+]_1}{r} = 264 \text{ mM}$$

(b) We can approximate the osmotic pressure difference by adding up the contributions of protein, potassium, and chloride:

$$\Delta\pi = RT \left\{ \Delta[P] + \Delta[K^+] + \Delta[Cl^-] \right\}$$

The various concentrations are

$$[K^+]_2 = 264 \text{ mM} \qquad\qquad [K^+]_1 = 300 \text{ mM}$$

$$[Cl^-]_2 = 342 \text{ mM} \qquad\qquad [Cl^-]_1 = 300 \text{ mM}$$

$$[P] \quad = \quad 10 \text{ mM}$$

in which $[Cl^-]_2$ has been found from the Donnan ratio. (Notice that charge in the protein compartment is conserved, within the present approximations). The osmotic pressure is, then

$$\pi_2 - \pi_1 = RT [10 + 42 - 36] \, 10^{-3} \, M$$

$$= .083 \, \frac{\text{lit-atm}}{°\text{K-mole}} \, 300° \, K \, (16) \, \frac{\text{moles}}{\text{lit}}$$

$$= 398.4 \times 10^{-3} \text{ atm.}$$

(c) At equilibrium

$$\Psi_2 - \Psi_1 = \frac{RT}{zF} \ln \frac{[K^+]_1}{[K^+]_2} = \frac{RT}{zF} \ln \frac{[Cl^-]_2}{[Cl^-]_1}$$

$$= 8.3 \, \frac{\text{volt-coul}(300)}{°\text{K mole } 96,500 \text{ coul/mole}} \ln \frac{264}{300}$$

$$= - 3.3 \text{ mV}$$

6-4. This problem has been discussed in the text with reference to the Nernst equation.

6-5. See Sec. 6-4 for calculation of the charge separated to obtain a given potential difference using the capacitance of the nerve membrane. The minimum work needed to recover these ions—if no other process were taking place—would be the electrochemical work

$$W_{\text{rev}} = n \, \Delta\tilde{\mu}$$

in which n is the charge separated across the membrane (expressed in moles of K^+) and $\tilde{\mu}$ the total electrochemical potential for K^+ in this system,

$$RT \ln \frac{K_{\text{in}}}{K_{\text{out}}} + F(\Psi_{\text{in}} - \Psi_{\text{out}}).$$

6-6. Clearly, a two compartment system can be established in which the membrane separates a reference solution (at known concentration of H^+) from the solution whose H^+ concentration must be measured. As the "membrane" is permeable only to H ions, the electrical potential measured across is a Nernst potential given by

$$\Psi - \Psi_{ref} = \frac{RT}{F} \ln \frac{[H]}{H_{ref}} .$$

The potential difference may then be calibrated to read the ratio of unknown to known H concentration or, since the known concentration is constant, the unknown concentration directly. In practice the "membrane" is a special ion-selective glass and one of the "compartments" is the barrel of an electrode made with the glass. The potential difference is measured between the reference electrode and the outside solution where the electrode is placed. This is the principle of the commercial pH meters.

6-7. Suppose that we can measure the potential on the surface of the nerve (which is much easier than measuring the potential inside the nerve and it can be done with large metal electrodes). The potential relative to the inside of the nerve is

$$\Psi_{out} - \Psi_{in} = \frac{RT}{F} \ln[K]_{out} - \frac{RT}{F} \ln[K]_{in}$$

if the Nernst Equation holds. If we plot the change in potential outside the nerve (relative to an arbitrary ground potential, for example) vs. the logarithm of external potassium concentration (which may be easily changed), a straight line should be obtained. This experiment, however, is not definitive because the reference potential inside—second term in the equation above—could have a different form and the same result would be obtained. But it can tell us if the Nernst Equation *is not obeyed*; hence, if potassium is not in equilibrium across the nerve membrane. Of course, the same measurement could be done for any other ion.

6-8. At $311°$ K $(38°C)$ the expression RT/F that appears in the Nernst Equation has a value

$$\frac{RT}{F} = \frac{8.3 \text{ joules}/° \text{ K-mole } 311° \text{ K}}{96,500 \text{ coul/mole}}$$

Since a joule is equal to 1 volt-coulomb, the final units are volts:

$$\frac{RT}{F} = 27 \text{ millivolts (mV)}.$$

The Nernst potentials for each of the ions, if they were in equilibrium would be (measured inside minus outside)

$$Na^+ : 27 \text{ mV } \ln\left(\frac{15}{150}\right) = -62 \text{ mV}$$

$$K^+ : 27 \text{ mV } \ln\left(\frac{150}{5.5}\right) = 89 \text{ mV}$$

$$Cl^- : 27 \text{ mV } \ln\left(\frac{9}{125}\right) = 70 \text{ mV}$$

The only ion that is in equilibrium is, then, Cl^-.

Chapter 7

7-1. (a) The flow J_+ is given by

$$J_+ = v_+ \, c$$

$$= \omega_+ \frac{\Delta\tilde{\mu}_+}{d} c$$

(b) as in (a),

$$J_- = \omega_- \frac{\Delta\tilde{\mu}_-}{d} c$$

(c) Since $\Delta\mu = RT \dfrac{\Delta c}{c} + zF\Delta\Psi$, we can write, for small Δc,

$$J_+ = \frac{\omega^+}{d} cR \frac{T\Delta c}{c} + zF \, \Delta\Psi$$

$$= \frac{\omega_+ RT}{d} \Delta c + \frac{\omega_+ c}{d} F\Delta\Psi$$

and

$$J_- = \frac{\omega_- RT\Delta c}{d} - \frac{\omega_- c}{d} F \, \Delta\Psi$$

7-2. (a) Let c_2 be the higher concentration. In that case, the logarithm in the formula

$$\Psi_2 - \Psi_1 = \frac{RT}{F} \frac{(\omega_- - \omega_+)}{(\omega_- + \omega_+)} \ln \frac{c_2}{c_1}$$

is positive; however, since the mobility of negative ions is larger than the mobility of positive ions, the overall expression is negative. The potential on the concentrated side is then negative relative to the dilute compartment.

(b) The concentrated compartment is positive.

(c) In this case, we see from the formula given above that $\Psi_2 = \Psi_1$.

(d) As charge is conserved,

$$J_+ = J_-$$

and

$$\frac{\omega_+}{d} RT\, \Delta c + \omega_+ \frac{c}{d} F\, \Delta\Psi = \omega_-\, RT\, \frac{\Delta c}{d} \overset{-\omega-}{} \frac{c}{d} F\, \Delta\Psi$$

Rearranging terms, we obtain

$$\Delta\Psi = \frac{RT}{F} \frac{(\omega_- - \omega_+)}{(\omega_- + \omega_+)} \frac{\Delta c}{c}$$

This expression may be given explicitly in terms of concentration ratios and potential differences as

$$\Psi_2 - \Psi_1 = \frac{RT}{F} \frac{(\omega_- - \omega_+)}{(\omega_- + \omega_+)} \ln \frac{c_2}{c_1}$$

(e) Using the formula for $\Delta\Psi$ and the expression for J_+ obtained in part (d)

$$J_+ = \omega_+ R \frac{T\Delta c}{d} + \omega_+ \frac{cRT}{cd} \frac{(\omega_- - \omega_+)}{(\omega_- + \omega_+)} \Delta c$$

$$= \omega_+ \frac{RT}{d} \frac{2\omega_-}{\omega_- + \omega_+} \Delta c$$

7-3. From the result obtained in Prob. 7-2 (c), it follows that a concentrated uni-univalent electrolyte in which the mobilities of anion and cation are equal does not introduce additional potentials such as those considered in 7-2.

7-4. The concentrations that appear in Fick's Law are expressed as a function of x. The difference in concentration is expressed as the concentration in the right side of the membrane minus the concentration on the left. If the concentration difference is positive (larger on the right than on the left), the flow will be negative—it will move in the direction of decreasing x, from right to left.

7-5. When n moles of the charged species are transported against the electrochemical potential difference, the work done is:

$$W = n\, \Delta\hat{\mu},$$

which can be equated to the work of displacement, $F \cdot d$, to yield $F \cdot d = n\, \Delta\hat{\mu}$

$$\frac{F}{n} = \frac{\Delta\hat{\mu}}{d}.$$

7-6. (a) The total number of flips is given by

$$N = H + T,$$

in which H is the total number of heads and T is the total number of tails.

(b) Since D increases or decreases one step at a time, D_N, the difference between the number of heads and tails after the Nth flipping, can have one of two possible values

$$D_N = D_N + 1$$

or
$$D_N = D_N - 1$$

(c)
$$(D_{N-1} + 1)^2 = D_{N-1}^2 + 2 + 1$$

and
$$(D_{N-1} - 1)^2 = D_{N-1}^2 - 2 + 1.$$

Over many trials both answers appear with equal frequency; the average displacement can be given by adding both results and dividing by 2:

$$D_N^2 \text{ (av)} = D_{N-1}^2 + 1$$

(d)
$$D_1^2 \text{ (av)} = 1$$

since
$$D_0 \quad = 0.$$

$$D_2^2 \text{ (av)} = D_1^2 + 1 = 2$$

$$D_3^2 \text{ (av)} = D_2^2 + 1 = 2 + 1 = 3$$

Clearly,
$$D_N^2 \text{ (av)} = N$$

(e) From the results above, it follows that

$$D_N \text{ (av)} = \sqrt{N}$$

The ratio of the displacement (average) to the total number of events, N, is then given by

$$\frac{D_N \text{ (av)}}{N} = \frac{\sqrt{N}}{N} = \frac{1}{\sqrt{N}}$$

7-7. The fact that the water flow is zero when the concentration difference is zero and then follows the same type of linear relation as passive flow, indicates that addition of the hormone leads to some structural change in the membrane which gives move effective area for the water molecules to go through, but no active transport.

Chapter 8

8-1. The information content of the English message calculated on the basis of all letters being equally frequent is

$$H = -\ln_2 p$$

$$= -\ln_2 \left(\tfrac{1}{26}\right)$$

$$= -\ln_2 26$$

In order to calculate the logarithm on the base 2 we can write

$$2^H = 26$$

and, taking natural logarithms on both sides,

$$H \ln 2 = \ln 26$$

$$H = \frac{\ln 26}{\ln 2} = \frac{3.26}{0.693}$$

$$= 4.7.$$

Note that the information content when all the probabilities are equal is higher than the information content with unequal probabilities (this is the same behavior shown by entropy).

8-2. If all the triplets are equally frequent, the probably of appearance of any triplet is

$$p(\text{AUG}) = p(\text{AAA}) = \ldots = \tfrac{1}{64}$$

The probability for each amino acid is, then, the sum of the probabilities of all triplets which can code for the amino acid; for example,

$$p(\text{Phe}) = p(\text{UUU}) + p(\text{UUC}) = \tfrac{2}{64}.$$

Similarly, we obtain

$$p(\text{Met}) = p(\text{Try}) = \tfrac{1}{64}$$

$$p(\text{Tyr}) = p(\text{His}) = p(\text{Gln}) = p(\text{Asp}) = p(\text{Lys}) = p(\text{Asp}) = p(\text{Glu}) = p(\text{Cys}) = \tfrac{2}{64}$$

$$p(\text{Ile}) = \tfrac{3}{64}$$

$$p(\text{Val}) = p(\text{Pro}) = p(\text{Thr}) = (\text{Ala}) = p(\text{Gly}) = \tfrac{4}{64}$$

$$p(\text{Leu}) = p(\text{Ser}) = p(\text{Arg}) = \tfrac{6}{64}$$

The information content calculated on the basis of amino acid messages is, neglecting chain termination signals,

$$-H = \Sigma p \ln p = 2(\tfrac{1}{64}) \ln_2(\tfrac{1}{64}) + 9(\tfrac{2}{64}) \ln_2(\tfrac{2}{64})$$

$$+ 1(\tfrac{3}{64}) \ln_2(\tfrac{3}{64}) + 5(\tfrac{4}{64}) \ln_2(\tfrac{4}{64}) + 3(\tfrac{6}{64}) \ln_2(\tfrac{6}{64})$$

$$= -4$$

If all amino acids are equally likely to occur, the information content is

$$H = -\ln_2(\tfrac{1}{20}) = 4.4.$$

8-3. The probability of finding each of the RNA nucleotides is:

$$p(\text{A}) = 0.878$$

$$p(\text{U}) = 0.105$$

$$p(C) = 0.014$$

$$p(G) = 0.003$$

The probability of appearance of each triplet can be calculated by assuming that each triplet is independent of neighboring triplets and that the probability of a nucleotide is not affected by the nucleotide which precedes it or follows. As an example, we calculate the probability of the sequence AUC,

$$p(AUC) = p(C)p(A)p(U)$$

$$= 0.878 \times 0.105 \times 0.014$$

$$= 0.00129$$

Clearly, this will also be the probability of appearance of the triplets CAU, ACU, CUA, UAC and UCA as we have assumed that nucleotides do not affect the neighboring nucleotides. Other probabilities are:

$p(A,U,G) = 0.00028$	$p(U,U,C) = 0.00015$
$p(U,C,G) = 0.00000$	$p(U,U,G) = 0.00003$
$p(A,A,U) = 0.08094$	$p(C,C,A) = 0.00018$
$p(A,A,C) = 0.01079$	$p(C,C,G) = 0.00000$
$p(A,A,G) = 0.00231$	$p(C,C,U) = 0.00002$
$p(G,G,A) = 0.00000$	$p(A,A,A) = 0.67683$
$p(G,G,U) = 0.00000$	$p(U,U,U) = 0.00115$
$p(G,G,C) = 0.00000$	$p(G,G,G) = 0.00000$
$p(C,G,A) = 0.00004$	$p(C,C,C) = 0.00000$
$p(U,U,A) = 0.00968$	

The probability of appearance of each amino acid is found by adding the probability of all triplets which stand for the amino acid (or punctuation mark); e.g.,

$$p(\text{Phe}) = p(UUU) + p(UUC) = 0.00115 + 0.00015$$

$$= 0.00130$$

Similarly,

$p(\text{Leu}) = 0.01117$	$p(\text{Pro}) = 0.00019$
$p(\text{Ile}) = 0.09191$	$p(\text{Thr}) = 0.01430$
$p(\text{Met}) = 0.00028$	$p(\text{Ala}) = 0.00004$
$p(\text{Val}) = 0.00031$	$p(\text{Thy}) = 0.01197$
$p(\text{Ser}) = 0.00178$	$p(\text{term}) = 0.08122$

$p(\text{His}) = 0.00147$ $p(\text{Asp}) = 0.00032$

$p(\text{GluN}) = 0.01083$ $p(\text{Cys}) = 0.00003$

$p(\text{AspN}) = 0.09093$ $p(\text{Try}) = 0.00028$

$p(\text{Lys}) = 0.67914$ $p(\text{Arg}) = 0.00235$

$p(\text{Glu}) = 0.00231$ $p(\text{Gly}) = 0.00000$

Although we have kept five figures after the decimal point to show how the probabilities are obtained, it is reasonable to retain only three (this gives a total sum of probabilities of 0.999 which is very close to 1). The information content is

$$H = - \Sigma p \ln_2 p = 2 \times 0.001 \ln_2 .001 + 2 \times 0.011 \ln_2 0.011$$

$$+ 2 \times 0.092 \ln_2 0.092 + 3 \times 0.002 \ln_2 0.002 + 0.014 \ln_2 0.014$$

$$+ 0.012 \ln_2 0.012 + 0.081 \ln_2 0.081 + 0.679 \ln_2 0.679$$

In this expression we have put together terms with the same probability. Also, note that terms with zero probability drop out because they can no longer be considered messages. The information content obtained this way is $H = 1.70$.

If the thirteen messages which comprise the final message were equally likely ($p = \frac{1}{13}$), the information content would be

$$H = - \ln_2 (\tfrac{1}{13}) = 3.71.$$

8-4. We can calculate the free-energy change required during reversible work of transport under the condition described,

$$\Delta G = RT \ln \frac{c_1}{c_2}$$

$$= RT \ln 2$$

$$= T \Delta_i S$$

As

$$H = \frac{S}{k}$$

we obtain

$$H = \frac{8.3 \times 6.9 \times 10^{-17} \text{ ergs/molecule}}{1.38 \times 10^{-16} \text{ ergs/molecule}}$$

$$= 2.87 \text{ bits.}$$

8-5. The error made by the enzyme is calculated as the fraction of the time the wrong substrate is recognized. For the first step we can calculate the error as

$$e_i = \frac{\text{valine loaded as Aacyl-AMP}}{\text{leucine loaded as Aacyl-AMP}} = \frac{13.2}{358} = 0.037$$

The overall error is

$$e = \frac{0.18}{3.2} = 0.056$$

As the probabilities of sequential events multiply, the intermediate, second-step error alone must be

$$e_2 = \frac{e}{e_1} = 1.525$$

(Strictly speaking this is a fractional error and not a probability.) A parallel error is added, rather than multiplied, so small sequential errors reduce the overall error, whereas small parallel errors may give a large total error.

Chapter 10

10-1. (a) During the two reversible adiabatic expansions the entropy change is zero. During the reversible compression at constant temperatures T_1 and T_0 the entropy changes are:

$$\Delta S_1 = \frac{Q_{\text{gained from bath at } T_1}}{T_1} = \frac{Q_1}{T_1}$$

and

$$\Delta S_2 = \frac{Q_{\text{lost to bath at } T_0}}{T_0} = \frac{-Q_0}{T_0}.$$

(b) The heat gained and lost during the processes given above are

$$Q_1 = T_1 \, \Delta S_1$$

and

$$Q_0 = -T_0 \, \Delta S_0,$$

respectively,

(c)

$$E_f = \frac{W_{\text{out}}}{Q_1}.$$

As W_{out} = heat gained - heat lost,

$$E_f = \frac{T_1 \, \Delta S_1 - T_0 \, \Delta S_0}{T_1 \, \Delta S_1}.$$

As the entropy, S, is a function of state, its value does not change after a full cycle; thus,

$$\Delta S_{\text{cycle}} = \Delta S_1 - \Delta S_0 = 0$$

$$E_f = \frac{T_1 - T_0}{T_1}$$

10-2. If $37°$ C ($310°$ K) were the high temperature, T_1, the external temperature would have to be, for a 20% efficiency,

$$\frac{310 - T_0}{310} = 0.2$$

$$T_0 = 310 \times 0.8 = 248° \text{ K} = -25° \text{ C}.$$

If, on the other hand, $37°$ C were the low temperature T_0, T_1 would be

$$T_1 = 0.2 \, T_1 + T_0$$

$$T_1 = \frac{T_0}{0.8} = \frac{310° \text{ K}}{0.8}$$

$$= 388° \text{ K} = 115° \text{ C}$$

(and all tissues would cook).

10-3. (a) $$\text{Power} = \frac{W}{t} = \frac{Fd}{t} = \frac{\Delta\mu}{d} v$$

(b) $$\frac{W}{t} = F\frac{d}{t} = PA\frac{d}{t} = P\frac{V}{t} = J_V P$$

(c) $$\text{Power} = \frac{W}{t} = F\frac{d}{t} = \left(\frac{F}{q}\right)\left(\frac{qd}{t}\right) = E\,l.$$

10-4. From the equality

$$\Delta G = -T\,\Delta_i S$$

we obtain, after dividing both terms by Δt,

$$\frac{\Delta G}{\Delta t} = \frac{T\Delta_i S}{\Delta t}.$$

10-5. The rate of internal entropy production is

$$\frac{\Delta_i S}{\Delta t} = \frac{(JX)}{T}$$

$$= (0.1 \text{ moles/sec})\left(\frac{RT \ln c_2}{c_1}\right) T$$

$$= (0.1 \text{ moles/sec})(1.99 \text{ Cal/degree mole } \ln 2)$$

$$= 0.199 \ln 2 \text{ cal/degree sec}$$

$$= 0.139 \text{ cal/degree sec}$$

10-6. The increase in entropy of the system during three seconds is

$$\Delta_i S = 3 \times 0.139 \text{ cal/degree} = 0.417 \text{ cal/degree}.$$

10-7. We define the positive flow as the one going from left to right and its conjugate force the potential difference between the left and right compartments. The dissipation function is, then,

$$\Phi = 0.025 \times RT \ln \frac{0.2}{1} + 0.01 \, RT \ln \frac{0.1}{0.3}$$

$$= T \times 0.025 \times 1.99 \text{ cal/degree mole moles/sec cm}^2 \, (-1.61)$$

$$= T \times 0.01 \text{ moles/sec cm}^2 \, 1.99 \text{ cal/degree mole } (-1.10)$$

$$= -(0.08 + 0.02) \, T = -.1 \, T.$$

Since the dissipation function is negative, the process is driven by the position dissipation provided by some other process. It is not passive.

10-8. The concentration of the filtrate is an indication of the relative volumes of solute to water that go through the membrane per unit time, hence the ratio of the solute to water velocities:

$$[glc]_{\text{filtrate}} = \frac{v_{glc}}{v_{H_2O}} \, [glc]_{\text{compartment}}$$

$$\frac{v_{glc}}{v_{H_2O}} = \frac{0.025}{0.1} = 0.25$$

As

$$1 - \sigma = \frac{v_{glc}}{v_{H_2O}}$$

we obtain

$$\sigma = 0.75.$$

The water velocity at zero hydrostatic pressure is given by

$$v_w = -\sigma L_p \, RT \, \Delta c$$

$$= -(0.75) \left(0.79 \times 10^{-6} \frac{\text{cm}^3}{\text{dyne-sec}}\right) \left(8.3 \times 10^7 \frac{\text{dyne-cm}}{\text{K}^\circ \text{ mole}}\right)$$

$$\left(300^\circ \text{ K}\right) \left(\frac{0.1 \text{ mole}}{1000 \text{ cm}^3}\right) \stackrel{\sim}{=} 0.15 \text{ cm/sec.}$$

and the volume flow through a 1 cm^2 membrane is then 1.5 cm^3/sec.

Appendix

A-1. (a) The First Law may be written in differential form as

$$dE = dQ - dW.$$

For work of expansion $dW = PdV$; introducing this expression and the temperature dependence of the energy change we obtain

$$C_V \, dT = dQ - PdV.$$

(b) The differential dQ/T can now be given as

$$\frac{dQ}{T} = C_V \frac{dT}{T} + \frac{P}{T} dV.$$

(c) If dS dQ/T is an exact differential,

$$\frac{\partial}{\partial V}\left(\frac{C_V}{T}\right)_T = \frac{\partial}{\partial T}\left(\frac{P}{T}\right)_V.$$

Since C_V is assumed to be a constant, the term on the left is zero. According to the equation of state for ideal gases

$$\frac{P}{T} = \frac{nR}{V}$$

so the second partial derivative is

$$\frac{\partial}{\partial T}\left(\frac{P}{T}\right)_V = \frac{\partial}{\partial T}\left(\frac{nR}{V}\right)_V.$$

As n and R are constants, this partial derivative is also zero, and we conclude that dS is an exact differential—S is a function of state.

A-2. (a) The change in entropy can be calculated by integrating dS by parts:

$$\Delta S = \int dS = \int_{T_0}^{T_1} \frac{C_V}{T} dT \bigg|_{\text{const.} V_0} + \int_{V_0}^{V_1} \frac{nR}{V} dV \bigg|_{\text{const.} T_1}$$

$$= C_V \ln \frac{T_1}{T_0} + nR \ln \frac{V_1}{V_0}.$$

We may also integrate the expression in parts for a second path

$$\Delta S = \int_{V_0}^{V_1} \frac{nR}{V} dV \bigg|_{\text{const.} T_0} + \int_{T_0}^{T_1} \frac{C_V}{T} dT \bigg|_{\text{const.} V_1}$$

$$= C_V \ln \frac{T_1}{T_0} + nR \ln \frac{V_1}{V_0}.$$

Clearly, the two integrations lead to the same result even though the paths are different.

A-3. (a) From $A = ST - E$ we obtain the differential

$$dA = S\,dT + T\,dS - dE$$

or, recalling the First Law,

$$dA = S\,dT + T\,dS - T\,dS + P\,dV$$

$$= S\,dT + P\,dV.$$

(b) Since A is a function of state, dA is an exact differential and it follows that

$$\left(\frac{\partial S}{\partial V}\right)_T = \left(\frac{\partial P}{\partial T}\right)_V.$$

(c) We wish to show that

$$\left(\frac{\partial E}{\partial V}\right)_T = 0.$$

From the First Law of Thermodynamics used in the case of reversible $P - V$ work,

$$dE = T\,dS - P\,dV;$$

it follows that

$$\left(\frac{\partial E}{\partial V}\right)_T = T\left(\frac{\partial S}{\partial V}\right)_T - P.$$

But, we showed in (b) that

$$\left(\frac{\partial S}{\partial V}\right)_T = \left(\frac{\partial P}{\partial V}\right)_V,$$

which, for an ideal gas is simply

$$\frac{\partial}{\partial T}\left(\frac{nRT}{V}\right)_V = \frac{nR}{V}.$$

Thus,

$$\left(\frac{\partial E}{\partial V}\right)_T = T\frac{nR}{V} - P = P - P = 0.$$

A-4. Suppose that two adiabats could actually cross in the $P - V$ plane. It would then be possible to set up a process in which a gas expands at constant temperature then compresses along an adiabat and expands along another adiabat to return to the original state. Heat would be gained from the constant temperature bath, and none would be lost to a lower temperature reservoir. All heat is converted into work, and this is impossible according to the Second Law of Thermodynamics.

A-5. From the First Law of Thermodynamics we have

$$dE = dq - dw.$$

For an ideal gas doing $P - V$ work this expression becomes

$$C_V\,dT = dq - P\,dV$$

and, as the expansion is adiabatic, $dq = 0$. The work done by the gas is simply given by the change in energy,

$$C_V \, dT = -dw;$$

furthermore, C_V can be assumed to be a constant, so the work is simply

$$-dw = C_V(T_2 - T_1).$$

The problem reduces, then, to finding what the change in temperature is when the volume expands from V_0 to $2V_0$ under adiabatic conditions. Introducing the equation of state for idea gases, $PV = nRT$ into the First Law expression with $dq = 0$, we obtain

$$C_V \, dT = nR \, \frac{T}{V} \, dV$$

or, rearranging,

$$\frac{C_V \, dT}{nRT} = - \frac{dV}{V}.$$

Integrating both sides we obtain

$$\frac{T_2}{T_1} = \left(\frac{V_0}{2V_0} \right)^{nR/C_V}$$

The work done by the gas is, then,

$$-w = C_V (T_2 - T_1) = C_V \left(\left(\frac{1}{2} \right)^{nR/C_V} - 1 \right) T_1.$$

AUTHOR INDEX

Agranoff, B. W., 199
Andersen, H., 169
Blout, E. R., 88
Boltzman, L., 68
Brillouin, L., 184, 192
Capp, A., 240
Carnot, S., 42, 219
Clausius, R., 42
Curran, P., 132
Dayhoff, M. O., 213
Diesel, R., 271
Dyson, F. J., 80
Eccles, J. C., 112
Eddington, A., 43
Edelman, G. M., 93
Einstein, A., 147
Eliassen, R., 249
Emden, R., 41
Frisch, K. von, 202
Gates, D. M., 4, 236
Gatlin, L., 188
Gessner, F., 126
Gibbs, J. W., 69
Irvine, E. C., 194
Joule, J. P., 36, 219
Katchalsky, A., 132, 231
Katz, B., 116
Kelvin, Lord, 42, 192
Kettlewell, H. B. D., 251
Kline, O., 126
Kormondy, E. J., 239

Landis, E., 131
Leaf, A., 165
Lehninger, A., 104, 105
McElroy, W. D., 227
Margalef, R., 252
Maxwell, J. C., 42, 68, 192
Mayer, J., 35
Miller, G., 197
Miller, S., 209
Morowitz, H., 209
Nash, L. K., 25
Odum, E. P., 237, 253
Odum, H. T., 236
Onsager, L., 228
Oparin, A., 208
Oplatka, A., 231
Pasteur, L., 2
Pimentel, N., 74
Prosser, C. L., 135
Quarton, C. G., 93
Quastler, H., 207
Rumford, Count, 35
Schmidt Nielsen, K., 136, 163, 238
Schramm, N., 126
Schrödinger, E., 2, 153
Sellinger, H. H., 227
Shannon, C., 177
Solomon, A. K., 165
Srb, A. M., 252
Starling, E., 131
Swanson, C. P., 159

Tanford, C., 146
Thompson, B. *See* Rumford, Count
Tinker, J., 208
Tribus, M., 194
Urey, H., 209
Ussing, H., 160
Villee, C., 16
Waddington, C., 204

Wald, G., 210
Watson, J. D., 154
Watt, J., 30
White, E. H., 69
Whittam, R., 103
Wiener, N., 182
Wilson, E., 201

SUBJECT INDEX

Absolute temperature. *See* Temperature
Action potentials, 114
Activation energies, 84
and enzymes, 85
of weak bonds, 86
Activity coefficients, 117
Adiabatic systems, 21
ATP:
in fermentation, 105
generation in biological systems, 102
in Krebs cycle, 105
and the recovery of chemical energy, 97
Autotrophs, 16
Avogadro's number, 62

Bees, communication in, 201
Biogeochemical cycles, 16
Biophysical thermodynamics, 1
Boltzman. *See* Maxwell-Boltzman distribution
Boltzman's equation, 72
Bonds:
activation energy of weak bonds, 92
covalent, 86
hydrogen, 90
weak, 90

Caloric theory, 35
Carnivores, 6
Cell, 10
membrane, 14
organelles, 13
Cellular specializations, 14
Chemical energy, recovery. *See* ATP
Chemical equilibrium, 265
Chemical potentials. *See* Potentials
Chemical reactions, free energy change, 81, 265
Chemomechanical coupling:
in an artificial system, 221
and asymmetry, 231
dissipation function, 229
Chlorophyll, 5
Chloroplasts, 14
Closed systems, 21
Code:
binary, 185
genetic, 180
information theory, 178
Morse, 178
Optimum, 182
Coding:
biological mistakes in, 189
in crayfish photoreceptors, 215
Codons, mRNA, 179
Communication:
in bees, 201

Communication (cont.)
  problem, elements of, 178
  telegraph as example of, 178
Concentration, 57
Conjugate flows and forces, 228
Consumers, in food chains, 16
Covalent bonds, 89
Cycle, Krebs, 105
Cycles, biogeochemical, 16
Cyclic processes, 26

Derivatives, partial, 256
Differential definition of chemical potential, 261
Differentials, exact, 256
Diffusion:
  exchange, 161
  Fick's law, 145
  gas, 156
  intuitive meaning of Fick's law, 146
  and microvilli, 159
  molecular model, 148
  and random walk, 148
  in respiration, 157
  and size of organisms, 150
Dissipation function, 266

Ecosystems, 15
Efficiency:
  of living organisms, 246
  and power, 244
  of producers and consumers, 239
  technical, conversion and application, 242
Elements:
  in the earth's crust, 208
  in the human body, 208
Energy, activation, 84
Energy:
  flow in ecosystems, 17
  free; see Free energy
  as function of state, 27, 37
  functions; see Gibbs free energy;
    Helmoltz free energy; Enthalpy
  human communication and, 195

  photosynthetic, 5
  solar, 3, 5
  transducers, 226
  units, 38
Energy metabolism, possible evolution of, 212
Engine:
  Carnot, 221
  efficiency of biological, 219
Enthalpy, 259
Entropy, 23, 42
  change during irreversible processes, 46
  as function of state, 47
  and probability, 70
Entropy production:
  internal, 46
  minimum, 252
  and spontaneous chemical coupling, 230
Enzymes, activation energies of, 85
Equilibrium, 22
Exact differentials, 256
Exchange diffusion, 161
Extensive variables. See Thermodynamic variables

Fermentation and ATP, 105
Fick's law of diffusion, 145
  differential formulation, 264
First Law of Thermodynamics, 259
Flows, conjugate, and forces, 228
Flows of mass in ecosystems, 17
Fluctuations:
  molecular, 154
  and size of organisms, 154
Food chains, 235
  and energy pyramids, 237
Force:
  conservative, 262
  driving, 140, 263
  and intensive variables, 141
Free energy, 49
  change, example, 54
  and chemical equilibrium, 81
  and chemical potentials, 108

and chemical reactions, 265
and Second Law of Thermodynamics, 50
and useful work, 51

Gas, ideal, 66, 67
Gas diffusion in respiration, 156
Genetic map of *E. coli*, 205
Genetic message, 178
Gibbs-Donnan equilibrium, red cells, 121
Gibbs-Duhem relation, 269
Gibbs free energy, 259
Glass electrodes and pH measurement, 118
Glucose oxidation, free energy change, 95

Heat:
   electric vs. gas, 244
   mechanical equivalent of, 36
   metabolic, 35
   production and body weight, 238
Helmoltz free energy, 259
Herbivores, 6
Heterotrophs, 16
Hormones, 10, 201
Hydrogen bonds, 90

Information content:
   of bees dances, 202
   and entropy, 193
   of fire ant trail, 203
   and probability, 186
Information storage in the brain, 198
Information theory:
   and coding, 178, 185
   and knowledge, 194
   Maxwell's demon, 192
   and memory, 197
   and protein evolution, 213
   and thermodynamics, 192
Intensive variables. *See* thermodynamic variables
Isolated systems, 21

Kinetic energy, 33
Kinetics of the ideal gas, 68
Krebs cycle and ATP, 105

Life:
   chemical probabilities, 209
   spontaneous emergence, 207
Light emission by the firefly, 227
Logarithmic curves, 58

Matter, microscopic models, 66
Maxwell demon, 192
Maxwell-Boltzman distribution, 76
Membrane capacitance, 112
   phenomenological coefficients, 167
Memory:
   and information theory, 197
   and macromolecules, 198
Metabolism, 7
Microscopic separation of charge, 112
Microvilli and diffusion, 159
Mitochondrion, 13
Mobility, 143
Mole, 57
Molecular biology, 180
Molecules, biological, 12
Motion of charge, 140
   average displacement in random molecular, 149
   Brownian, 147
   and extensive variables, 141
mRNA codons, 179

Nephron, 162
Nucleus, cell, 14

Open systems, 21
Organisms:
   unicellular, 150
   volume to area ratios, 152
Osmotic equilibrium and chemical potentials, 123
Osmotic gradients, interaction with hydrostatic gradients, 165

Osmotic pressure, values for salt solutions, 130
Osmotic pressure in capillaries, 130
Osmotic regulation and habitat, 134

Permeability and lipid solubility in plant cells, 166
Phenomenological equations, 229
    proper, 268
Pheromones, and insect communication, 201
Photosynthesis, 5
Poiseulle flow, 164
Pollution:
    and species organization, 249, 251
    thermodynamics and environmental, 241
Pores in membranes, 164
Pores in red cells, 165
Potential energy, gravitational, 34
Potentials:
    action, 114
    chemical, 51, 59
    chemical, derivation, 260
    electrical, 51
    electrochemical, 260
    Nernst equation, 110
    in nerves, 111
    in neurons, 112
Power, 224
Pressure:
    of ideal gas, 76
    turgor, 125
Processes:
    cyclic, 26
    irreversible, 42
    reversible, 42
Producers, 16
Production, primary energy, 230

Redundancy:
    in biological systems, 188
    in coding, 188
Reflection coefficients, 131, 169
    measurements, 134

Respiration, 6
Ribosomes, 14

Second Law of Thermodynamics, 42
    and energy quality, 80
    and probability, 70
Sickle cell disease, 189
State:
    equation of, 24
    functions of, 25
    steady. *See* Steady states
Statistical mechanics, 69
Steady states, 140
    and final velocity, 142
    and local equilibrium, 222
Stomatal opening, 125
Systems, thermodynamic, 19

Temperature, 20
    absolute, 19
Thermodynamic variables:
    extensive, 23
    intensive, 23
Thermodynamics:
    biological, 1
    First Law of, 27, 259
    Second Law of; *see* Second Law of Thermodynamics
Thermostatics, 25
Trancriptase, reverse, 180
Translation, genetic, 180
Transmitters, irreversible, 194
Transport:
    active, 160
    facilitated, 160
    and kidney function, 163
    work, 140

Velocity, final:
    and Newton's Second Law, 142
    relative, of solvent and solute, 172
    solvent in dilute solutions, 167
Volume, partial molar, 60, 270

Work:
  biological, 6, 10, 94
  of expansion, 31

mechanical, 28
reversible, 256
of transport, 140